The Sun Today

Claudio Vita-Finzi

The Sun Today

 Springer

Claudio Vita-Finzi
Department of Earth Sciences
Natural History Museum
London, UK

ISBN 978-3-030-04078-9   ISBN 978-3-030-04079-6   (eBook)
https://doi.org/10.1007/978-3-030-04079-6

Library of Congress Control Number: 2018961191

Cover illustration:
The Daniel K. Inouye (Advanced Technology) Solar Telescope has a 4 m diameter primary mirror. It
is sited on the summit of Haleakala ('House of the Sun') on the Hawaiian island of Maui. Viewed
against the Milky Way galaxy. Photo by Scott Lacasse

This Springer imprint is published by the registered company Springer Nature Switzerland AG
The registered company address is: Gewerbestrasse 11, 6330 Cham, Switzerland

For Ennio

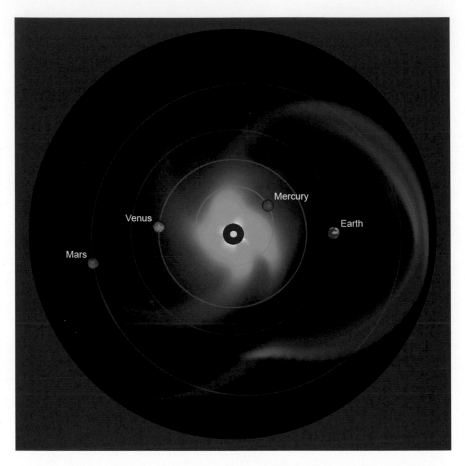

Frame from NASA movie of a model for the track of a fast coronal mass ejection of March 15, 2013 when it has crossed the inner solar system. Courtesy of Goddard Media Studios, NASA

Preface

The world's a scene of changes; and to be
Constant, in Nature were inconstancy.

Abraham Cowley (1618–1667)

By now, thanks to reports on TV and in magazines, we are all aware that the Sun is a moody creature subject to tantrums and unpredictable sulks, and warnings about the damage one of its explosions could cause our satellites and power supplies sound increasingly urgent. On the other hand, the possible role of the Sun in global warming and in precipitating droughts and famines tends to be eclipsed, for reasons good and bad, by explanations which highlight human folly. Add to the mix conflicting reports about the benefits and potential harm of exposure to the Sun's rays and we are left with a poor grasp of what we could call the solar factor.

Three of my Springer books have touched on the history of the Sun. The present book focuses on its current nature; a set of stills from an array of films. If to a moviegoer 'A stopped frame outside of a movie isn't anything, not even a photograph'[1], to a student of nature it is just as helpful as were Muybridge's images of galloping horses: witness the solar storm snatched from a NASA movie (Frontispiece). Even so, some flexibility is needed to capture a mood, especially as the Sun of today embodies events that occurred elsewhere in the solar system aeons ago. Perhaps it is better to think of this book as more like a TV interview, where some reminiscence is expected but the central character remains in view.

Previous studies of the Sun's activity and its interaction with us on Earth include both highly technical accounts and treatments aimed at the non-scientific reader[2]. My book is neither fish nor fowl (I'd call it a monograph if it were not so weedy[3]): it summarises what we have learnt from recent satellite missions, observation from the Earth, and the fruits of theory and computer modelling, and it also proposes a novel mechanism for heating the corona to 2 million K and for structuring the solar atmosphere. But its intended audience is ill-defined. To echo John Bahcall[4], the book aims to 'share the fun of figuring out some of nature's puzzles'; and surely half the fun is learning new stuff: popular science which excessively dilutes the science cheats the reader just as the language of the Reader's Digest Condensed Books series evaporated away some of the books' literary qualities while safeguarding their narratives. At all events, the text of The Sun Today is self-contained but extensive reference lists will enable the reader to follow up points of interest in the library or on the Internet and they allow me to give credit for the many ideas and data I cite.

Loren W. Acton, Edward Hanna, John Marshall and Gary J. Rottman generously commented on the text. I am again indebted to the space agencies, and above all NASA, for access to the fruits of their missions, to Michael Woolfson, Judith Lean, Ken Phillips and Leo Vita-Finzi for stimulating discussions, to Petra van Steenbergen and Hermine Vloemans at Springer for support, to Simon Tapper for help with the figures, and to Scott Lacasse for the cover image.

London, UK   Claudio Vita-Finzi
September 2018

References

1. M Wood (2012) *Film*. Oxford Univ. Press, New York
2. Examples: A Bhatnagar & W Livingston *Fundamentals of solar astronomy* (2005); H Friedman *Sun and Earth* (1985); L Golub & J M Pasachoff *Nearest star* (2nd ed 2014); K R Lang *The Sun from space* (2000); D J Mullan *Physics of the Sun* (2010); K J H Phillips *Guide to the Sun* (1992); C P Sonnett et al. (eds) *The Sun in time* (1991); M Stix *The Sun* (2nd ed 2002); O R White (ed) *The solar output and its variation* (1977); H Zirin *Astrophysics of the Sun* (1988).
3. The Minigraph: the future of the monograph? stunlaw.blogspot.com/2013/08
4. J N Bahcall (1989) *Neutrino Astrophysics*. Cambridge Univ. Press, Cambridge

Abbreviations

AU = Astronomical unit = Average distance from Earth to Sun = $\sim 150 \times 10^6$ km.
ly = Light year = $\sim 9.5 \times 10^{12}$ km.
pc = Parsec = ~ 3.26 ly = 3×10^{13} km.
$M_\odot$ $L_\odot$ $R_\odot$ = Solar mass, solar luminosity, solar radius.
Gyr = 10^9 yr Myr = 10^6 yr.

Some of the numerous acronyms that infect the subject are spelled out in the text or in the index.

Contents

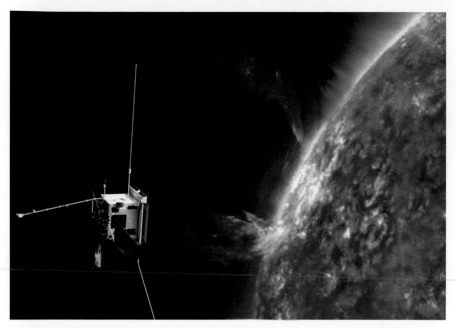

ESA's next-generation Sun explorer, Solar Orbiter, will be launched in 2020. It will investigate the connections and the coupling between the Sun and the heliosphere. Image credit ESA/AOES

Chapter 1
A Commonplace Star

Abstract In 1911 the Sun shed its unique role in humanity's universe and joined the ranks of several million G2V yellow dwarf stars; whatever the resulting loss of prestige by association, humanity has thereby gained greatly in its understanding of solar history and probable future by analogy with kindred stars and the calibration of models of solar evolution. The Sun's composition and inner workings were revealed by spectroscopy and nuclear physics; advances in solar physics and chemistry illuminate other stars, solar systems and galaxies, to the benefit of our cosmic understanding. Numerous devices now monitor the Sun, their development prompted by necessity as well as curiosity.

The Sun gained the status of star whenever astronomers accepted a multiplicity of solar systems. Democritus (~460–370 BC), for example, referred to innumerable other worlds some of which had no sun and moon and others more than one, and Aristarchus of Samos (~310–230 BC) suggested that the stars were distant suns[18]. The motion bubbled to the surface from time to time thereafter even though it must have seemed a confusing if not heretical diversion from the business of understanding our immediate planetary environment. In the 16th century it was voiced by Bruno, in the 17th by Descartes and von Guericke, and in the 18th by William Herschel and Immanuel Kant, whose nebular hypothesis (1755) envisaged a stage when multiple suns revolved around a galactic centre.

Readers familiar with the achievements of modern astronomy may already feel impatient with a historical digression, but the roots of the subject are intriguingly hardy. The main categories of stars are still based on the classification of Hipparchus (~130 BC); the droll dwarf-giant notation that was introduced over a century ago has survived[7, 25]; and even the stellar temperature calibration introduced by Annie Jump Cannon in 1918–1924 remains largely in place. Today's solar astronomy tolerates such archaisms, like poultices in a transplant clinic, while it promotes and exploits great advances in hardware and ideas.

© Springer Nature Switzerland AG 2018

C. Vita-Finzi, *The Sun Today*, https://doi.org/10.1007/978-3-030-04079-6_1

A G2V Star

An early move towards reform, or at any rate formalisation, of solar physics as a whole took place a century ago thanks to the marriage of spectroscopy and nuclear physics. The former had been rendered astronomically valuable by the identification of spectral lines with specific elements by Robert Bunsen and Gustav Kirchhoff (Fig. 1.1); nuclear physics soon impinged on astronomy by advances in atomic theory and in particular the notion of ionization embodied in an equation published by Meghnad Saha in 1920 which linked the ionization state of a gas (that is, the extent to which its electrons had been ejected) to the prevailing temperature and pressure. The formula, as Cecilia Payne[24] was quick to realise, rendered it immensely influential in the analysis of stellar atmospheres.

A related development hinged on the comparison between the brightness or luminosity of the visible stars and their spectral type. A plot by Ejnar Hertzsprung[13] of the apparent magnitude of the Hyades cluster against their colour as indicated by their effective wavelength in Ångstroms, and thus their inferred temperature, revealed a pattern which was dominated by a diagonal belt (Fig. 1.2a). At about the same time Henry Norris Russell[25] was working on a similar concept, with spectral class on the horizontal axis and visual magnitude (to be precise, the magnitude that the star would have if brought to the distance of 10 parsecs from Earth = 32.6 light years) on the vertical axis. When the graph was constructed (Fig. 1.2b) the spectrum of over one hundred thousand stars had been determined but the parallax of only 300 had been measured directly. The value of parallax, that is the apparent shift in the relative position of an object in response to a shift in the location of the observer, is of course required for establishing distance and thus for converting m to absolute magnitude.

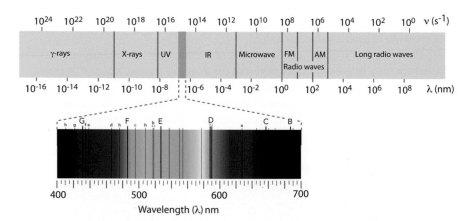

Fig. 1.1 Visible part of the electromagnetic spectrum (380–710 nm) and Fraunhofer lines (courtesy of Wikimedia Commons)

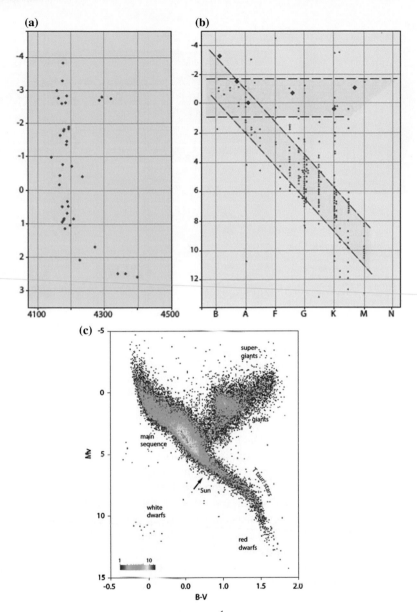

Fig. 1.2 **a** Plot of magnitude against wavelength in Ångstroms for Hyades (based on Hertzsprung[19], Fig. 7). **b** Plot of absolute magnitude against spectral type (Harvard Observatory scale) after Russell[33]. The heavy dots represent stars observed once; the smaller dots represent stars with parallax based on the mean of two or more determinations. **c** Hertzsprung-Russell (HR) diagram for the 41,704 single stars from the Hipparcos Catalogue. Colours indicate the number of stars. The stellar temperatures are represented by a colour index, B-V, and brightness is given as the absolute magnitude in the Hipparcos photometric system. To reveal the relations between temperature and brightness the measurements of absolute magnitude are shown as if the source were observed at a distance of 10 parsec. Courtesy of ESA. Conventional star nomenclature added

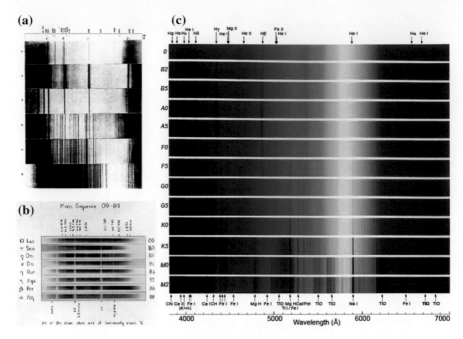

Fig. 1.3 a The Harvard classification of stellar spectra (from Charlier[3]). **b** Spectra for Main Sequence O9-B9 stars (from Morgan et al.[21]). **c** Stellar spectra. Note how the spectra of colder stars (F/G/K/M: lower rows) are more complex than those of hotter stellar atmospheres (O/B/A: upper rows), which are dominated by H and He (courtesy of NASA, Planetary Spectral Generator)

Hertzsprung-Russell (or HR) diagrams (Fig. 1.2c), as they came to be called, are sometimes said to herald the birth of modern astronomy. As colour photography had not yet been developed the spectra on which they hinged were entirely monochrome and thus the emphasis was on line position and thickness (Fig. 1.3a, b). The spectral categories were devised by Cannon and others at the Harvard Observatory in the 1920s on the basis of naked eye inspection of a small part of the spectrum[30]. In the Morgan-Keenan variant of the scheme each of the categories O-M that appears on Fig. 1.3c is qualified from hottest to coolest by the addition of digits 0–9, and divided into luminosity classes from O to V, with Roman V standing for main-sequence stars. Various other refinements have been introduced but the core principle has proved durable.

The inverse scale for classifying stellar brightness arose because for Hipparchus the brightest stars were termed 'of the first category', the faintest 'of the sixth category' with four grades in between. The apparent magnitude (m) of a star as seen from the Earth is the inverse of its brightness, so that the Sun has a m of -27 and the stars at the limit of visibility by naked eye have a m of 6–7. It presages the L-shaped pattern of modern Hertzsprung-Russell figures, with the horizontal axis decreasing in value away from the vertical axis (e.g. Fig. 1.2c)

Luminosity assessment, however, was perforce subjective. Russell noted that the luminosity of stars at the top of the diagram was 7500× that of the Sun, and the luminosity of one at the bottom of the diagram 1/5000 of the Sun's, and he recognised two great classes of stars, one on average 100× brighter than the Sun and one of lesser brightness, which he called respectively giant and dwarf stars. These terms had been introduced by Hertzsprung in 1905 for the reddest stars on the Harvard 1901 catalogue (K and M on Fig. 1.3a) according to whether they were brighter or less bright than our Sun; they persist even though they are also applied to groupings (such as white dwarfs) very different from the original range of categories.

Russell commented that, if we could put on his diagram some thousands of stars instead of the 300 at his disposal, and further reduce the uncertainty over their absolute magnitude, they would cluster along the B-M diagonal and also along a second line also starting at B but nearly horizontal. The modern composite H-R diagrams are dominated by the diagonal ('main-sequence') grouping as well as a number of groupings off it, with various alternative scales for the two axes, including temperature vs luminosity or colour (spectral type) vs absolute magnitude. The plot in Fig. 1.2c contains 41,704 stars derived by the Gaia UK project from the Hipparcos satellite (1989–1993), a catalogue[17] of over 1 billion stars brighter than magnitude 20.7.

Sun-like Stars

Setting aside any nostalgia for a unique Sun to power a unique planet, the individuality of our host star needs to be reassessed from time to time as science progresses, because the outcome will colour any extrapolation to or from other planetary systems and also any conclusions derived from ground-based and satellite observation of star composition and behaviour. Thus some accounts of the search for extraterrestrial life, and the reconstruction of terrestrial climate history, assume that life could not survive the temperatures resulting from an increase in solar luminosity[12] by a factor of about 2, while calculations suggest that planetary systems which include inhabitable terrestrial planets are most probable around stars of one stellar mass $(M_\odot)^{34}$.

In a workshop held in 1998 the issue was discussed with reference to mass, age, chemical composition, angular momentum, differential rotation, granulation and turbulence, activity and binarity (membership of a binary system)[9]. The discovery of numerous solar systems has rendered suzerainty over a planetary retinue commonplace; and in many other respects the Sun appeared unremarkable, though somewhat richer in iron than most coeval stars. However, a reassessment in 2008 concluded that the Sun is peculiar in some respects when compared with stars of similar age and galactic orbit. It is more massive than most 'solar twins', stars with main parameters almost identical to the Sun's as compared with solar analogues and solar-type stars. Its activity as defined at visual wavelengths varies less than similar stars; it is a single star whereas most solar-type stars are thought to form part of binary or multiple systems; it possesses an unusual set of planets; and it seems to be poorer than they are in refractory elements, that is to say those resistant to heat[10].

The effect could be explained by conditions in the protoplanetary nebula in which the solar system originated when the formation of planetesimals and terrestrial planets had depleted its refractory ('dust') component[9, 11]. One implication would then be that other stars with compositions similar to the Sun's are found to host terrestrial planets, one example being HIP 102152 which is viewed as a 'twin' of the Sun though ~3.6 Gyr older[20]. Thus the question of uniqueness may appear frivolous, but it takes on special significance in the search for extraterrestrial life. If the chemical composition of the Sun is unusual—so the argument runs—and in a manner that makes it especially favourable to life, say by the prominence of a particular element, then the hunt could focus on Sun/planet systems with similar characteristics.

'Similar galactic orbit' sounds too abstruse a criterion for identifying solar twins: but even from the viewpoint of present-day observation the orbit colours many aspects of solar behaviour. The Sun is in the Orion Spur between the Sagittarius and Perseus Arms of the Milky Way barred galaxy, which is composed of 200 billion stars, has a diameter of about 100,000 ly, and is within the 'Local Group' of about 36 galaxies (Fig. 1.4a, b). The Spur rotates around the galactic centre at 217 km/s taking ~226 Myr to complete a circuit. Two other motions—an oscillation towards and away from the galaxy centre at 20 km/s and another oscillation at 5–7 km/s above and below the plane of the galaxy—would seem even less relevant to our current concerns, yet they now enter into discussion of the mechanisms of climate change, as once did zones of cosmic radiation and dust clouds[14, 32].

To be sure, the timescale in question is expressed in millions of years and the oscillations were invoked to explain the periodicity of ice ages, also measured in millions of years (e.g.[29]), but the mix of human and cosmic time scales is a useful reminder that the two are bound to overlap from time to time. One such coincidence is the flux of galactic cosmic rays (GCRs), which have long been linked to weather[22] and whose level fluctuates measurably from minute to minute (Fig. 1.5).

What do other stars tell us? We have seen that Sun-like stars may shed light on the Sun's probable past and possible future, but they also complement solar data with regard to present-day processes such as the workings and location of the dynamo[23] by providing a population whose mass and rotation can be viewed statistically. The temptation must have been strong from the outset to assume that stars on the main sequence would share an evolutionary history: the word 'sequence' implies it, as do phrases such as 'once a main sequence star has burnt through the hydrogen in its core it reaches the end of its life cycle. It now leaves the main sequence.' There is thus the bold assumption that all the stars on the MS are powered by the conversion of hydrogen into helium.

The observations that led to the first Hertzsprung- Russell diagrams exploited the expansion of distant stellar images by spectral analysis. Telescopic inspection of stars, even those in the galactic vicinity, has since gained greatly in locational precision but is still limited to various spectroscopic ruses. Whether the target is coming or going relative to the observer ('radial velocity') can be gauged by combining spectral data with the Doppler effect, and rotation can be measured by applying this technique to opposite edges of the star. For distant stars, where the sides of the star are indistinguishable and the star is spinning quickly we rely on rotational broadening,

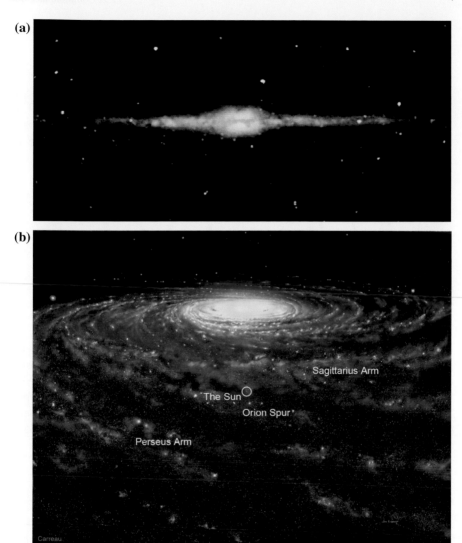

Fig. 1.4 a The Milky Way galaxy viewed by Cosmic Background Explorer (COBE) from the vicinity of our Sun at infrared wavelengths, which penetrated the dust and gas that would otherwise obscure the galactic centre (courtesy of NASA, COsmic Background Explorer (COBE) Project). **b** Environs of the Sun in the Milky Way galaxy: artist's impression (courtesy of ESA: © C. Carreau)

where light is recorded from all parts of the star simultaneously and individual lines are broadened.

Such esoteric matters as the temperature gradient from the photosphere through the chromosphere into the corona (Fig. 1.6) or the operation of flares are inferred by spectral subtraction, using a synthesised stellar spectrum[15]. Asteroseismology,

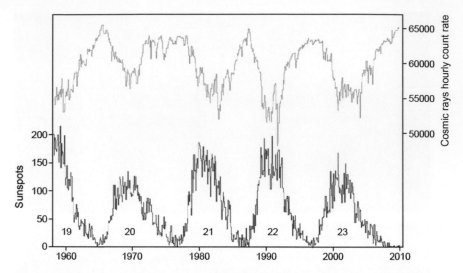

Fig. 1.5 Galactic Cosmic Rays count rate and sunspot number for Cycles 19–23 illustrating the Forbush effect (courtesy of Dr. Ole Humlum)

the determination of stellar interiors by reference to their oscillations viewed as the product of seismic waves, is done by combining observations of the luminosity, radial velocity and spectra of pulsating stars[15]. Even starspots and differential rotation can be inferred from photometry and colour indices, notably using the Kepler Space Telescope[36]. Conversely, solar properties underpin much work on extrasolar planetary systems both implicitly and by reference to standard solar models (SSMs: see Chap. 3)[31].

Observing the Sun

Today in the title of this book refers to the time of observation rather than of formation, on the understanding that it consequently lumps together items that came into being at any time between the Big Bang and the current (solar) year. In the cosmic context it amounts to microhistory, 'the mode of scholarly investigation that studies a specific moment, event, person, or thing in close, sometimes obsessive detail'[1].

The range of techniques for observing our Sun has multiplied in variety as well as precision. The key avenues of exploration have so far been rockets, balloons and artificial satellites and probes; the expansion of the wavelengths from the ground on either side of the visible range to include radio and infrared besides the well tried UV sources; magnetic field measurements; and direct sampling of the solar wind. As usual, curiosity-driven research yields unexpected and novel results. The chance observation of a solar flare by Richard Carrington[2] led indirectly and after many hiccups to the recognition of a Sun-Earth link and in effect to what we would now

Fig. 1.6 Schematic
cross-section of the solar
atmosphere

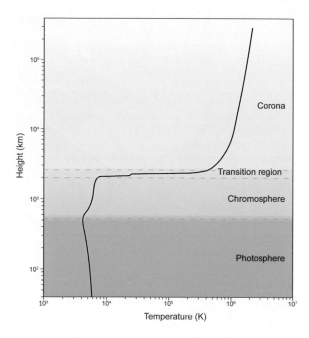

call space weather. And what at the time must have seemed little more than an oddity likely at most to inconvenience telegraph operators has now ballooned into a global concern thanks to the development of artificial satellites, sensitive electronics and the militarisation of space[33].

The opening salvo was fired in 1946 by V2 rockets observing the far ultraviolet (FUV). There soon followed orbiting solar observatory (OSO) missions (1962–1986) which gathered data on solar flares and X-ray radiation, followed by the Skylab mission (1973–1979; manned 1973–1974) and the Helios probes (1974 and 1974) at 0.3 AU, that is to say inside the orbit of Mercury. The Solar and Heliospheric Observatory (SOHO 1995-) has proved especially durable and productive: planned for 2 years it has performed for over 20. The Sunrise solar telescope, with its one-meter mirror the largest solar telescope to fly above the atmosphere, flew in 2000 at a height of about 30 km for five days to obtain detailed images of the chromosphere.

By 2008 some 6404 spacecraft starting with Sputnik 2 (1957) had been launched of which 188 were specifically concerned with solar physics, and this total does not include all the satellites associated with space physics (632) and astronomy in general (299) that bear on the Sun indirectly or accidentally (Fig. 1.7).

The ensuing years have witnessed a number of ambitious solar missions, notably the SDO (Solar Dynamics Observatory: 2010-), PICARD (2010–2014), named after the 17th century solar astronomer Jean Picard, IRIS (Interface Region Imaging Spectrograph: 2013-), and DSCOVR (Deep Space Climate Observatory: 2015-), and at the time of writing there are plans for probes which will closely approach the Sun in 2018, including the Solar Orbiter (to within 45 solar radii) and the Parker Solar

Fig. 1.7 The Helioseismic and Magnetic Imager (HMI) instrument on the Solar Dynamics Observatory (SDO) satellite (courtesy of NASA)

Probe (to 3 solar radii: Fig. 6.1). As with their predecessors, these missions have a multiplicity of aims. SDO, for example, was devised to gather data bearing on the Sun's 11-year cycle of activity, the development of magnetic active regions on the Sun, magnetic reconnection, variations in the Sun's EUV output, the role of magnetic fields in coronal mass ejections, the solar wind, and space weather. Ulysses (1990–2008) occupied an unusual orbit as, unlike all its predecessors, it passed over the solar poles in order to fulfill its task of imaging the Sun at all latitudes.

Since the 1920s observation of the Sun from the ground or space, like that of other solar system bodies[9], has depended heavily on spectroscopy, although the range of wavelengths, the quality of the spectra and the analytical procedures have of course greatly improved over the years. For earthbound solar observatories the scope for spectral diversification is limited by atmospheric absorption of X-rays, much of the solar UV radiation and infrared radiation. The classification of stellar spectra that was developed at the Harvard Observatory by Cannon ordered stars in terms of their photospheric temperature. The range of wavelengths observed within which line strengths were evaluated for this purpose is 380–750 nm, although in practice much seems to have been accomplished in the 390–470 nm segment. It has been suggested that the success of radio astronomy demonstrated that there was much to be learnt outside the optical range and thus spurred on the development of infrared astronomy [2009]. Whatever its initial motivation the diversification continues. The working solar beam of the Inouye telescope on Haleakala volcano in Hawaii, for example (cover), will be split so that various instruments can observe simultaneously the same, overlapping or different wavelength bands from near UV to Far IR.

In addition, modern solar telescopes devote much ingenuity to reducing atmospheric turbulence outside and within the telescope itself. To this end the 1.6 m primary mirror in the McMath-Pierce telescope (1962) at Kitt Peak is housed underground; in the Inouye the 4.2 primary mirror will employ adaptive optics powered by computer-controlled pneumatic and hydraulic devices; stellar spectroscopy as a whole has embraced an extreme version of adaptive optics which promises to achieve radial velocity precisions measured in metres per second[16].

However, direct measurement of extreme ultraviolet (EUV), an important ingredient of space weather research, is impossible from the ground because of atmospheric absorption, and none of the surrogates that are otherwise used, such as the sunspot number, is entirely satisfactory[4]. The same applies to soft X-rays, part of the radio range and gamma rays. Besides circumventing the atmospheric barrier, observation from space platforms may also provide observing conditions that cannot be attained on Earth[8], such as the binocular viewpoints occupied by the STEREO spacecraft and, of course, the polar orbits of Ulysses.

Improvements in resolution characterise space observation no less than terrestrial devices. Take 'magnetic reconnection', a mechanism much favoured as a source of coronal heating (see Chap. 5) but hitherto incompletely imaged. In 2013 the High Resolution Coronal Imager on a sounding rocket photographed the Sun and captured imagery at the resolution required to investigate the reconnection process. The results owed much to the optical imaging system, which had a resolution 14 times better than that of the SDO.

The improvement is increasingly digital rather than optical. In 2008 an international consortium produced an automated system for the SDO for identifying almost in real time selected features in the 1.5 TB of data delivered daily by its instruments and where appropriate—as in the case of solar flares—tracking them in order to forecast their likely development[19], with obvious implications for space climate and encouraging a shift in solar physics towards 'helioinformatics'.

As discussed in Chap. 3 the solar wind samples the corona and photosphere and thence the Sun's outer convection zone; the usual assumption is that this mirrors the parent nebula of 4.6 Gyr ago though with some alterations due to general gravitational settling and selective chemical changes in the Sun and in the solar wind itself[35]. Comparison between the solar wind and spectral measurements of solar composition is thus a window into processes which have been operating in the Sun since it accreted.

But this is to ignore the value of the solar wind as solar breathalyser. Two major sources on the Sun for this afflatus are the coronal 'holes' first observed from Skylab (Fig. 1.8), which yield the fast solar wind, and part of the Sun's equatorial belt. Coronal holes serve as windows into the lower corona or perhaps even the photosphere whereas the fast solar wind samples a mixture of sources. The proton record at numerous observatories (Fig. 1.9) provides an indication of gross changes in solar wind energy and thus its value as shield against the flux of galactic cosmic rays.

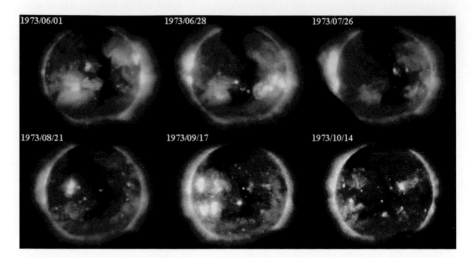

Fig. 1.8 Evolving coronal 'hole', where plasma density is very low, imaged by Skylab in soft X-rays (0.12–5.0 nm) in June–October 1973 (courtesy of NASA)

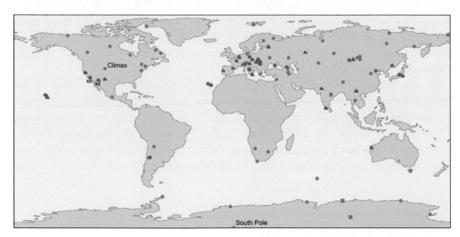

Fig. 1.9 Earthbound observatories. Blue triangles show astronomical observatories, green blobs show neutron monitors

References

1. Bugg J (2017) On the origin of faeces (review). Times Lit Suppl 27 oct no 5978, p. 33
2. Carrington RC (1859) Description of a singular appearance seen in the Sun on September 1, 1859. Mon Not Roy Astron Soc 20:13–15
3. Charlier CVL (1921) Lectures on stellar statistics. Scientia, Lund
4. Dudok de Wit T, Kretzschmar M, Aboudarham J, Amblard P-O, Lilensten J (2008) Which solar EUV indices are best for reconstructing the solar EUV irradiance? Adv Space Res 42:903–911
5. Eddington AS (1914) Stellar movements and the structure of the Universe. Macmillan, London
6. Encrenaz T (2008) Remote sensing analysis of solar-system objects Phys ScrT130: 014037
7. Gingerich O (2013) The critical importance of Russell's diagram. arXiv:1302.0862v1 [physics.hist-ph]
8. Golub L (2003) Solar observation from space. Rev Sci Instr 74:4583
9. Gustafsson B (1998) Is the Sun a Sun-like star? Space Sci Rev 419–428
10. Gustafsson B (2008) Is the Sun unique as a star—and if so, why? Phys Scr T130: 01430
11. Gustafsson B, Meléndez J, Asplund M, Yong D (2010) The chemical composition of solar-type stars in comparison with that of the Sun. Astrophys Space Sci 328: 185–191
12. Handler G (2013) Asteroseismology. In: Oswalt TD et al (ed) Planets, stars and stellar systems, v 4. Springer, Netherlands, 207–241
13. Hertzsprung E (1911) Über die Verwendung photographischer effektiver Wellelängen zur Bestimmung von Farbenäquivalenten. Pub Astrophys Obs Potsdam 22:63
14. Hoyle F, Lyttleton RA (1950) Variations in solar radiation and the cause of ice ages. J Glaciol 1:453–455
15. Husser T-O et al (2013) A new extensive library of phoenix stellar atmospheres and synthetic spectra. AsrXiv:1303.5632v2 [astro-ph.SR]
16. Jovanovic N, Schwab C, Cvetojevic N, Guyon O, Martinache F (2016) Enhancing stellar spectroscopy with extreme adaptive optics and photonics. Pub Astr Soc Pacific 128:121001
17. Lindegren L et al (2016) Gaia data release 1. Astrometry: one billion positions, two million proper motions, and parallaxes. Astron Astrophys 595:A4
18. Lloyd GER (1996) Adversaries and authorities. Cambridge University Press, Cambridge
19. Martens PCH et al (2012) Computer vision for the *Solar Dynamics Observatory* (SDO). Solar Phys 275:79–113
20. Monroe, TR and 15 others (2013) High precision abundances of the old solar twin HIP 102152: insights on Li depletion from the oldest sun. ApJL 771:L31
21. Morgan WW, Keenan PC, Kellman F (1943) An atlas of stellar spectra with an outline of spectral classification. Astrophys Monog, Univ Chicago Press, Chicago Ill
22. Ney EP (1959) Cosmic radiation and the weather. Nature 183:451–452
23. Noyes RW, Baliunas SL, Guinan EF 1991 what can other stars tell us about the Sun? In Cox AN, Livingston WC, Matthews MS (eds) Solar interior and atmosphere, Univ Ariz Press, Tucson, 1161–1185
24. Payne CH (1925) Stellar atmospheres. Harvard Univ Monog, Cambridge Mass
25. Russell HN (1913) 'Giant' and 'dwarf' stars. Observ 36:324–329
26. Russell HN (1914) Relations between the spectra and other characteristics of the stars. Pop Astron 22:275–294
27. Saha MN (1921) On a physical theory of stellar spectra. Proc Roy Soc, https://doi.org/10.1098/rspa.1921.0029
28. Savanov IS, Dmitrienko ES (2017) Spots and activity of solar-type stars from Kepler observations. Astron Rep 61:461–467
29. Shaviv NJ, Prokoph A, Veizer J (2014) Is the Solar System's galactic motion imprinted in the Phanerozoic climate? Sci Rep 4, 6150
30. Sobel D (2016) The glass universe. 4th Estate, London
31. Turck-Chièze S (2016) The Standard Solar Model and beyond. Jour Phys, Conf Ser 665:012078
32. Vidal-Madjar A et al (1978) Is the solar system entering a nearby interstellar cloud? Astrophys J 223:589–600

33. Vita-Finzi C (2015) A perfect solar storm. Proc Am Phil Soc 159:1–6
34. Wetherill GW (1996) The formation and habitability of extra-solar planets. Icarus 119 219–238
35. Wiens RC, Bochsler P, Burnett DS, Wimmer-Schweingruber RF (2004) Solar and solar-wind
    isotopic compositions. Earth Planet Sci Lett 222:697–71

Chapter 2
An Inconstant Star

Abstract Stars have long been classified on the basis of their brightness; variability was a secondary consideration but eventually proved an important clue to stellar dynamics as well as a means of classification. The Sun's ~11-year activity cycle, identified mainly from sunspots and other surface features, is superimposed on both longer and shorter periodicities which are manifested in luminosity and internal processes to different degrees. Solar observation thus makes increasing demands on the versatility and sensitivity of observatories, observers and their archives, and rules out an all-purpose definition of the present-day Sun.

Recognition that stars may be impermanent could be said to date from the first recorded observation by Chinese astronomers of a nova in 1300 BC[30]; 'a great new star', it dwindled after two days. The supernova of AD 1054, of which the Crab Nebula in the constellation of Taurus is the residue (Fig. 2.1), was visible for 23 days. Many such 'guest stars' were recorded in China, although a few of them turned out to be comets or perhaps meteors. Variable stars were also known in mediaeval Japan and China to astronomers who, unlike their Arab and European counterparts, were not inhibited by 'prejudice and spiritual inertia'[30] in abandoning the notion of celestial perfection.

When it comes to the Sun many investigators assumed constant irradiance until 1980, when measurements by the Active Cavity Radiometer Irradiance Monitor I (ACRIM I) on the Solar Maximum Mission (SMM) satellite supplied incontrovertible evidence of solar inconstancy. Advances in satellite technology had made possible sustained and stable measurement outside the atmosphere, while developments in pyrheliometry—where the heat resulting from solar radiation is compared with that produced by a known amount of electrical power[34]—provided the required accuracy and consistency.

C. Vita-Finzi, *The Sun Today*, https://doi.org/10.1007/978-3-030-04079-6_2

Fig. 2.1 The Crab Nebula, in the Perseus Arm of the Milky Way 6500 ly from Earth, is the remnant of a supernova witnessed in China in AD 1054. Its radiowaves pass through the solar corona every June and help to define its extent

We said 'outside the atmosphere', but orbiting satellites have by no means bypassed the atmosphere entirely: the SMM satellite, for example, was brought down in 1989 by atmospheric drag resulting from increased solar activity in Cycle 22 (Fig. 1.5). A glowing UV geocorona (Fig. 2.2) produced by far-UV light (Lyman α) scattered by neutral hydrogen has been detected by satellites[31] at a height of 100,000 km.

The Solar Constant

The solar constant, a term long used rather loosely, is now defined[29] as the amount of solar radiation per unit area received on a normal plane at the top of the atmosphere

Fig. 2.2 Colour enhancement of a far UV (FUV) photo taken in 1972 from the Moon during Apollo 16 of the Earth's geocorona, the outer part of Earth's atmosphere which is composed of low density hydrogen. The Sun is to the left (courtesy of NASA)

at 1 AU (an astronomical unit, the average distance between Earth and Sun). It is sometimes used synonymously with total solar irradiance (TSI), where the pitfall of 'constant' is swapped with the ambition of 'total': for the range of wavelengths emitted by the Sun, as a later paragraph suggests, is incompletely known, and in any case, many instruments employed to measure solar variation, including ACRIM, do not identify the wavelengths in question[11] or use 'all wavelengths' to mean 'essentially the energetically important range from 200 to 5000 nm containing 99.9%' of the solar constant[12].

Early attempts to define the solar constant were less formal but often had the unstated aim of assessing the distortion of the Sun's irradiance produced by the Earth's atmosphere (Fig. 2.3) or by variations in the Earth-Sun distance[1, 23].

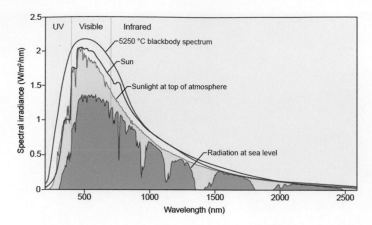

Fig. 2.3 The solar spectrum at the Sun, at the top of the atmosphere and at the Earth's surface compared with that of a 5250 °C blackbody (various sources, not always in agreement)

Instrumental measurement of these effects had begun in 1838, with Claude-Servais Pouillet's pyrheliometer (Fig. 2.4a). Pouillet defined the solar constant as the calorific strength of the Sun per cm^2 above the atmosphere, in contrast with the atmospheric constant, which embodies the variable level of transmission through the atmosphere. The result, which Pouillet expressed in calories/minute/cm^2, was 1228 W/m^2. The modern value is 1370 W/m^2, based on rather more complex equipment (Fig. 2.4b). S. P. Langley's determination on Mt Whitney, at an elevation of 4420 m, was 2140 W/m^2, and was accepted as the solar constant for over 20 years[10].

The need for prolonged data runs in order to identify trends and anomalies inevitably calls for the record of different instruments or missions to be fused and if necessary adjusted to avoid gaps or discontinuities. Lack of continuity has dogged the issue. The first of successive TSI measurements—the Nimbus 7/Earth Radiation Budget Experiment (1978–1993)—had an instrumental precision of 0.01% and an accuracy estimated at ±0.2% but it was primarily devoted to downward Earth observation rather than solar scrutiny. The Active Cavity Radiometer Irradiance Monitor (ACRIM I) on the Solar Maximum Mission (SMM, 1980–1989), with which the Nimbus7 mission overlapped, was Sun-pointing every two minutes and had three radiometers.

The complexity of the exercise is illustrated by the PREMOS/PICARD findings for 2011–2014, which revealed 'divergent trends' by other TSI instruments and prompted the conclusion that composites of TSI observations are still too uncertain to estimate TSI between solar cycle minima[7]. Yet, as the time-resolution of radiometers improves, so does our grasp of the extent to which the observed oscillations are real and not entirely a product of instrumental or atmospheric interference. Put more simply, although long-term stability can be assessed (it was better than 1 part in 10^5 per annum in the first 20 years), the absolute accuracy of the data was more difficult to estimate[13] and its significance to comprehend.

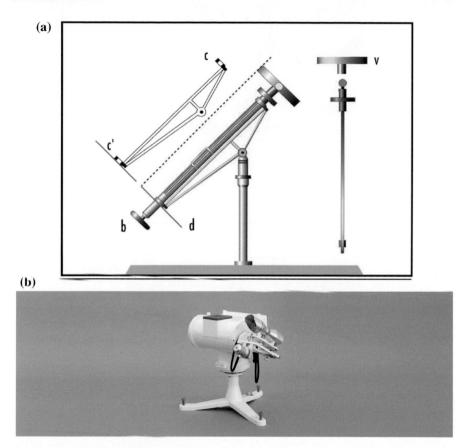

Fig. 2.4 **a** Pouillet's pyrheliometer (1837), which yielded a solar constant of 1.76 cal cm^{-2} min^{-1}. The current accepted value is 1.951 cal cm^{-2} min^{-1}. **b** MS-56 Pyrheliometer (2018), which measured direct normal incidence irradiance (DNI) and was sensitive to solar irradiance in the spectral range 200–4000 nm (image courtesy of EKO)

By 2005 it was still necessary to acknowledge that the (by then) 32-year TSI database, a composite derived from several overlapping spaceborne measurements, was marred by uncorrected instrumental drift and was too short and imprecise to establish any long-term changes in the irradiance[18].

A well documented example is the gap between the ACRIM I and ACRIM II sequences (June 1989 to October 1991) (Fig. 2.5), which was plugged with data from two other instrumental records as well as corrections of the data in the light of instrument degradation from exposure to solar radiation[14]. Note that this correction can be applied only if a less exposed radiometer of the same type is available, as with ACRIM; otherwise interpolation has to depend on appropriate modelling[14]. The gap was significant not only as a hiccup in a potential 25-year sequence but also because the choice of model would help to determine whether there was a long-term trend

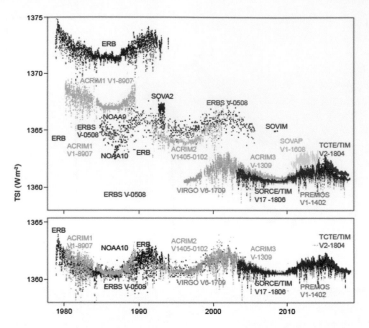

Fig. 2.5 Closing the two-year gap in the ACRIM record of solar radiation using overlapping TSI spacecraft missions, also allows differences in calibration between instruments to be corrected (courtesy of Greg Kopp)

between the minima for solar cycle 22 (1986) and solar cycle 23 (1996), a valuable complement to the search for cumulative change that would otherwise have to depend on averaging oscillations in the data. In the event, no evidence for an increase in TSI between those two dates emerged[19] but there is by no means unanimity in the choice of method use to bridge the gap.

The TSI is therefore a delicate, hybrid flower. Measurements made in 2008 using the Total Irradiance Monitor (TIM) on NASA's Solar Radiation and Climate Experiment (SORCE) satellite coupled with new laboratory tests, gave[18] the substantially lower result of 1360.8 ± 0.5 W m^{-2}. The measurements were made during the 2008 minimum (Fig. 1.5) but the crucial innovation was to limit extraneous scattered light from the detector, an interesting illustration of how scientific progress does not necessarily require the development of novel experimental devices or an intellectual revolution. Subtle statistical ruses[9] are employed not only to overcome data gaps but also to estimate the uncertainties. Indeed the acceptance of long-term (secular) changes in irradiance may require uncertainties as small as 0.01% and annual stabilities as small as 0.001% p.a. The data may repeatedly demonstrate the reality of the Schwabe (~11 year) solar cycle and its 22-year magnetic counterpart, but solar observation makes additional demands, such as repeated imaging at ever quicker tempo (cadence) in order to permit stringent testing of those physical models of solar activity that embody such oscillations or to detect short-term changes which might otherwise escape unnoticed.

Early moves in recognising solar variability were cautious; yet it is the interpretation that is nowadays sometimes premature, as where variation is explained by the rotation and distribution of sunspots[3] rather than anything more fundamental. Consider the range of amplitudes and periods so far recorded for some of the 43,675 variable stars that had been catalogued in the Milky Way by 2011. They range from 1/1000 of a stellar magnitude to 20 stellar magnitudes and the periods range from a fraction of a second to a number of years[4]. For the Cepheid variables the corresponding values are 2–60 days and 300–40,000 $L_\odot$.

And the slipperiness of all the criteria in question is illustrated by 'the evolutionary timescale' of white dwarf G117-B15A, termed 'the most precise optical clock known'. The star boasts six pulsation modes of which some vary in amplitude from night to night, while the periodicity of the main pulsation period is changing progressively at a rate that is consistent with the star's cooling timescale (namely 0.05 K/year for a core temperature of 1.2×10^7 K)[17]. Indeed it has been suggested that the pulsating variables which include Cepheids, RR Lyraes, Semiregulars and Miras may represent transitional periods between stages of evolution rather than permanent characteristics. In short one needs to distinguish long-term trends from short-term fluctuations.

Labelling the Sun a variable star must have at first seemed premature as, lèse-majesté apart, it was difficult to distinguish between intrinsic and extrinsic variability, that is to say between the star's variability and an effect produced by the atmosphere or some other external factor especially before the terrestrial atmosphere could be sidestepped—and indeed before timekeeping was technically up to the task. Clocks able to distinguish time intervals consistently, and thus able to identify astronomical periodicities, date from the mid 17th century with the development of Huygens' pendulum, coarse when compared with current resolutions measured in picoseconds[16] but benefiting from a substantial observation period (see White Dwarf G117-B15A discussed below).

Moreover, as data accumulate it is becoming clear that there are variations other than the Schwabe which could eventually emerge as true cycles. The solar constant is thus part of the complex set of changes some of which reinforce each other while others help to stifle their impact. The SOLSTICE (SOlar Stellar Irradiance Comparison Experiment) instrument on the UARS satellite from 1991 to 2001 provided long-term measurements of solar UV and FUV in order to improve understanding of solar radiation below 300 nm as this is completely absorbed by the atmosphere and becomes the main energy input into atmospheric processes[32]. A number of bright blue stars (O and B spectral type) known to vary by a small fraction of 1% over long periods provided a stable reference for tracing degradation of the SOLSTICE instrument.

The cadence of current instruments on the SDO satellite shows how technical progress can temporarily overtake known requirements and perhaps trigger new lines of research. A study of observations spanning eight years for 33 solar-type stars showed brightness changes year-to-year which were far greater than the equivalent for our Sun, perhaps by as much as a factor of four. The implication is that the Sun is going through an unusually steady phase[26].

Thus a succession of TSI measurements is evidently essential for pursuing such comparisons or more generally for tracing gross solar irradiance or modelling the future. It is also required for the more trivial aim of identifying a representative or anomalous year for a study such as the present. As we have seen, the 'history of solar irradiance determinations is, with few exceptions, the history of efforts to patch the infrequent or too discontinuous space-based measurements by devising mechanisms for proxy determination of solar radiance fluxes from ground-based obtained parameters'[8]. And we can maximise the available empirical data by judicious modelling, for example[35] by building on time series for facular brightening and sunspot darkening to model irradiance variability.

Analysis of full-disk magnetograms (Fig. 2.6) for the period since 1974 is claimed to reproduce up to 97% of the recorded irradiance variation[20]. Shorter-term variations, such as P mode oscillations (which peak at a period of ~5 min) or solar granulation may well be of 'no importance for climate studies'[20] but they are critical to the kind of portraiture being attempted here.

Spectral Effects

Assessment of the solar factor is hampered among other things by effects which distort the Sun's spectrum, especially in response to differential absorption and scattering by the atmosphere. For instance, there is a complete extinction of solar ultraviolet radiation (wavelengths shorter than about 300 nm) because of ozone and molecular oxygen absorption in the middle atmosphere. Outside the atmosphere the spectrum approximates that of a blackbody at 5772 K but even here it is interrupted by a number of dark absorption lines. Modelling to remedy the defects in the record include the SATIRE (Spectral and Total Irradiance Reconstructions) family of models. These are relevant to terrestrial climate, that is to say if they are on timescales longer than a day and appear to be entirely due to changes in the number and distribution of magnetic features on the solar surface.

We have seen that TSI is sometimes defined by wavelength though commonly in the visual and infrared range, just as were the early measures from aircraft, balloons, rockets and the early satellites. Modelling to remedy the defects in the observations include the SATIRE (Spectral and Total Irradiance Reconstructions) family of models. These bear on variations which, according to the modellers, are relevant to terrestrial climate because they are on timescales longer than a day and appear to be entirely due to changes in the number and distribution of magnetic features on the solar surface.

Fig. 2.6 Magnetogram for 18 December 2014. Two large sunspot groups with strong magnetic intensity stand out. The stronger black and white areas indicate more powerful polarity. These two active regions have unleashed a number of flares (courtesy of Solar Dynamics Observatory, NASA)

There is clearly something to be gained from a standard spectrum against which to assess distortions. A recent example dating from 2011 is the SCIAMACHY spectrum, which extends[15] from 235 to 2384 nm. The instruments of the ISS-SOLAR mission, part of the Columbus Laboratory mounted externally on the International Space Station (Fig. 2.7) and launched in 2008, in combination measure solar radiation at 17-3000 nm. One of the instruments, SOLSPEC, which operated until February 2017, measured the energy of each wavelength in absolute terms and also its variability so that a solar reference spectrum spanning 165–3000 nm and representative of the 2008 solar minimum was obtained with a resolution of better than 0.1 nm below 1000 nm and of 1 nm in the 1000–3000 nm range. Integrating the spectrum yields a

Fig. 2.7 SOLAR, an ESA observatory on the Columbus laboratory which forms part of the International Space Station. It carries three complementary instruments which make measurements of the Sun's spectral irradiance from 17 to 100 mm, within which 99% of the solar energy is thought to be emitted

total solar irradiance of 1372.3 ± 16.9 Wm $^{-2}$ at one sigma, i.e. 11 W m^{-2} over the value recommended by the International Astronomical Union in 2015[28]. These data are considered to embrace 99% of the Sun's irradiance.

Different parts of the spectrum vary over the Schwabe cycle—and not always in phase. This variability has to be specified when depicting the spectral composition of the 'present-day Sun' either with error values or by presenting the spectrum at solar maximum and minimum. Even then long-term variations have to be written in, with the date of observation specified as closely as possible so that the measurements can if necessary be corrected later. It has long been known that the UV portion of the solar spectrum changes in the medium-term much more than other wavelengths, and spectral irradiance variability is an order of magnitude larger at shorter ultraviolet wavelengths than in the visible and near infrared spectral regions[24] (Fig. 2.8). During a solar flare the Sun's extreme ultraviolet output can vary by factors of hundreds to thousands in a matter of seconds. Surges of EUV photons heat Earth's upper atmosphere, causing the atmosphere to "puff up" and drag down low-orbiting satellites. EUV rays also break apart atoms and molecules, creating a layer of ions in the upper

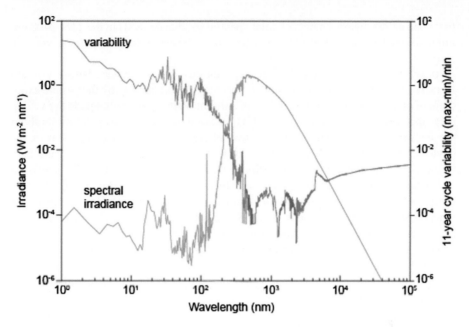

Fig. 2.8 Variability of solar spectral irradiance (grcen line, left hand scale) and differences in variability of different wavelengths (blue line, right hand scale). Note disproportionate variability of short wavelengths (courtesy of Judith Lean)

atmosphere that can severely disturb radio signals. EUV controls Earth's environment throughout the entire atmosphere above about 100 km; UV as a whole is seen as a key component in atmospheric heating.

As late as 1991 direct measurement of the spectrum above the atmosphere was limited to the few minutes allowed by rocket and balloon soundings, and this was supplemented by modelling atmospheric transmission to correct ground based spectral measurements, combining an intensity spectrum from a solar model with observation, and computing a theoretical flux spectrum from available data[21]. Matters changed with the launch of a number of spacecraft dedicated to the analysis of the solar spectrum. The motivation was to tease out those parts of the spectrum that changed most over time or that had the greatest influence on climate and health.

The VIRGO experiment, which rides on the SOHO spacecraft, provides since 1995 a time series for TSI and also solar irradiance at 402, 500 and 862 nm. Similarly, SORCE, which has monitored the Sun since 2003 and consists of four experiments including SOLSTICE (SOlar Stellar Irradiance Comparison Experiment), is designed to measure both total solar radiation and portions of the electromagnetic spectrum defined by X-ray, ultraviolet, visible, and near-infrared radiation. SORCE defined a reference spectrum of the Sun's irradiance from 0.1 to 2400 nm during very quiet solar conditions. Another major achievement of SORCE is that the satellites acquired the first continuous measurements of solar spectral irradiance (SSI) in the 115–2400 nm range as well as daily UV (115–320 nm) measurements for comparison with 18 stable stars.

In practice we are blinkered in our expectations, just as we were before William Herschel in 1800 recognised an infrared source of solar heating by the simple expedient of observing a rise in the reading of a thermometer placed alongside the visible solar spectrum. The definition of 'total' is coloured by reasonable assumptions about the impact of solar radiation on the atmosphere or some other component of the environment of human concern. However, to characterise the present day Sun there is every reason for exploiting the entire electromagnetic radiation (EMR) spectrum and indeed going beyond its conventional limits. Consider the disc of Galaxy NGC 4625 (in Canes Venatici), which was reported in 2018 to appears four times larger when viewed in UV light than in optical light, a phenomenon which could be telling us that many young, hot stars dated to a mere ~1 Gyr (compared with the estimated age of stars ~10 Gyr in the optical centre) are forming in its outer regions.

One might apply a similar comparative approach to solar imaging. The channels imaged by the Atmospheric Imaging Assembly (AIA) on the SDO satellite have an impressive range (Fig. 2.9) and thus document the solar atmosphere from the photosphere at 5000 K to flares[25] at 6.3 million K. However, although the radiation covered includes the UV and visible zones in their entirety and part of the X-ray and infrared regions, it omits any shorter or longer wavelengths. Nor does it exploit radar mapping, which is well established in terrestrial and planetary geology to complement optical imaging, a notable example being the Synthetic Aperture Radar (SAR) technique employed by the Magellan survey of Venus.

Radio astronomy, the passive counterpart to radar mapping, is more relevant to our case because it relies on radiation from the subject body itself. Indeed radio astronomy was born from the detection of cosmic radio noise from Sagittarius A at the centre of the Milky Way Galaxy. Microwave radiation from the Sun was first detected[33] in 1942–3. The Sun is one of the strongest radio sources in the sky (Fig. 2.10). It is primarily a thermal source characterised by high frequencies, which are thought to originate near the photosphere. A non-thermal component, represented by lower frequencies deriving from synchrotron radiation arising in the Sun's magnetic field, originates perhaps surprisingly in the corona[6]. The potential of radio techniques for investigating solar activity is shown by the observation, reminiscent of what we reported about Galaxy NGC 4625, that at frequencies between 0.1 GHz and 3 GHz the Sun is larger than its optical counterpart. Moreover, at wavelengths greater than 1 cm the flux density differentiates between a quiet Sun and a Sun displaying vigorous sunspot activity[6].

LOFAR, a low-frequency radio telescope array which links several north-west European countries and has a diameter of over 1000 km, is a project designed to trace processes in the corona at high temporal and spatial resolution as a contribution to the analysis of space weather. For example, the plasma frequency associated with coronal mass ejections (CMEs) in the low corona falls into LOFAR's frequency range (30–240 MHz)[5]. Type III radio bursts, which are due to electrons released from the

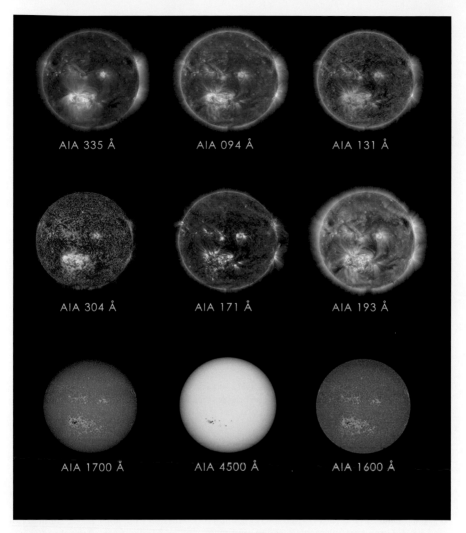

Fig. 2.9 Images of the Sun at various wavelengths by the Atmospheric Imaging Assembly (AIA) instrument on the Solar Dynamics Observatory spacecraft (courtesy of NASA and ESA)

Sun at the lift-off time of CMEs, may allow us to identify CMEs capable of creating powerful shocks which can accelerate charged particles to high speeds resulting in radiation storms[27].

At the short wavelength end of the spectrum, X-rays and gamma rays are both used in medical imaging and increasingly in astronomy. Gamma rays (energies > 100 keV, $\lambda < 10^{-11}$ m) have been ascribed to particle acceleration by solar flares[22]. In 2013 and 2014 the Fermi Gamma-ray Space Telescope traced gamma rays to flares associated with CMEs[2], some of them on the far side of the Sun.

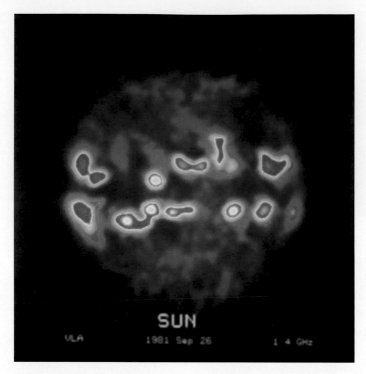

Fig. 2.10 Radio pictures of the Sun at 1.4 GHz recorded by the Very Large Array (VLA) of the National Radio Astronomy Observatory. The brightest features correspond to active regions (image courtesy of NRAO/AUI)

References

1. Abbott CG, Fowle FE Jr (1908) Recent determination of the solar constant of radiation. Jour Geophys Res 13:79–82
2. Ackermann M et al (2014) High-energy gamma-ray emission from solar flares: summary of *FERMI* Large Area Telescope detections and analysis of two M-class flares. Astrophys J 787:15
3. Albrect R, Maitzen HM, Rakos KD (1969) The Sun as a Variable Star. Astron Astrophys 3: 236–242
4. AAVSO (American As Variable Star Observers) (2017) www.aavso.or
5. Bastian T (2004) Low-frequency solar radiophysics with LOFAR and FASR. Planet Space Sci 52:1381–1389
6. Bradaschia F (2013) Radioastronomy. Sandit
7. Cessateur G et al (2016) Total Solar Irradiance changes between 2010 and 2014 from the PREcision MOnitor Sensor absolute radiometer (PREMOS/PICARD). AGU Fall Ass 2016, Abs SH42B
8. Domingo V (1994) In Pap JM et al (eds) The Sun as a variable star, IAU Colloq 143, Cambridge Univ P, Cambridge
9. Dudok de Wit T et al (2017) Methodology to create a new total solar irradiance record. Geophys Res Lett 44: 1196–1203 https://doi.org/10.1002/2016gl071866
10. Dufresne J-L (2008) La détermination de la constante solaire par Claude Pouillet. Météo 60:36–43

11. Frieman EA et al (ed) (1994) Solar influences on global change. Nat Acad Press, Washington DC
12. Fröhlich C (2016) Irradiance observations of the Sun. Intern Astron Un Colloq 143: 28–36
13. Fröhlich C & Anklin M (2000) Uncertainty of total solar irradiance: an assessment of the last twenty years of space radiometry. Metrologia 37:387–392
14. Fröhlich C & Lean J (1998) Total solar irradiance variations: The construction of a composite and its comparison with models. Proc Int Astr Un 185:89–102
15. Hilbig T et al (2016)The new SCIAMACHY reference solar spectral irradiance and its validation. Geophys Res Abs 18
16. Kalisz J (2004) Review of methods for time interval measurements with picosecond resolution. Metrologia 41:17–32
17. Kepler SO et al (2000) Evolutionary timescale of the pulsating white dwarf G117-B15A: the most stable optical clock known. Astrophys Jour 534: L185–L188
18. Kopp G, Lean JL (2011) A new, lower value of total solar irradiance: Evidence and climate significance. Geophys Res Lett 38: L01706, https://doi.org/10.1029/2010gl04577
19. Krivova NA, Solanki SK, Wenzler T (2009) ACRIM-gap and total solar irradiance revisited: is there a secular trend between 1986 and 1996? arXiv:0911.3817v1[astro-ph.SR]
20. Krivova NA, Solanki SK (2013) Models of solar total and spectral irradiance variability of relevance for climate studies, in Lübken F-J (ed) (2013) Climate and weather of the Sun-Earth system (CAWSES). Springer, 19–38
21. Kurucz RL (1991) The solar spectrum. In: Cox AN, Livingston WC, Matthews MS (eds) Solar interior and atmosphere. Univ Arizona Press, Tucson AZ, 663–669
22. Kuzhevskii BM (1982) Gamma astronomy of the Sun and study of solar cosmic rays. Soviet Phys Uspekhi 25:392–408
23. Langley SP (1884) Researches on solar heat and its absorption by the Earth's atmosphere. Rep Mount Whitney Exped, Prof Pap Signal Serv15. Washington
24. Lean J (2000) Evolution of the Sun's spectral irradiance since the Maunder Minimum. Geophys Res Lett 27:2425–2428
25. Lemen JR et al (2012) The Atmospheric Imaging Assembly (AIA) on the Solar Dynamics Observatory (SDO). Sol Phys 275:17–40
26. Lockwood GW, Skiff BA, Baliunas SL, Radick RR (1992) Long-term solar brightness changes estimated from a survey of Sun-like stars. Nature 360:653–655
27. Mäkelä P et al (2015) Estimating the height of CMEs associated with a major SEP event at the onset of the metric type II radio burst during solar cycles 23 and 24. Astrophys J 806:13
28. Meftah M et al (2017) SOLAR-ISS: a new reference spectrum based on SOLAR/SOLSPEC observations. Astron Astrophys doi.org/https://doi.org/10.1051/0004-6361/201731316
29. NASA (2017) www.mynasadata.larc.nasa.gov/
30. Needham J (1959) Science and civilisation in China, 3: Cambridge Univ Press, Cambridge
31. Østgaard N et al (2003) Neutral hydrogen density profiles derived from geocoronal imaging. J Geophys Res 108:A7s
32. Rottman GJ, Woods TN, McClintock W (2006) SORCE solar UV irradiance results. Adv Space Res 37:201–208
33. Southworth GC (1945) Microwave radiation from the Sun. J Frank Inst 239:285
34. Willson RC (1984) Measurements of solar total irradiance and its variability. Space Sci Rev 38: 203–242
35. Yeo KL, Krivova NA, Solanki SN (2017) EMPIRE: A robust empirical reconstruction of solar irradiance variability. arXiv: 1704.07652v1 [astro.ph.SR]

Chapter 3
Solar Composition

Abstract Until the late 1920s it was accepted that Sun and Earth had very similar compositions. The revelation that the Sun is composed primarily of hydrogen prompted novel models for its evolution and hence for solar irradiance and magnetism, and it was an essential step towards the current nuclear scheme with its dependence on hydrogen-helium transformation. Nowadays solar composition is investigated by a number of strategies which bear on different parts of the Sun, notably spectroscopy primarily of the photosphere and direct chemical assay of the corona by way of the solar wind, complemented by geochemical analysis of pristine carbonaceous chondritic meteorites, which are thought to have originated in the same nebula as the Sun. The results are evaluated in the light of models of the solar interior and the findings of helioseismology, and they bear on attempts to trace the origins of the solar system, the genesis of stars, and ultimately the origin of the elements in our galaxy and indeed in the universe as a whole.

Astronomers long assumed that the composition of the Sun was broadly similar to that of the Earth as they had both originated in the primaeval nebula that had hatched our solar system. The discovery in the 1920s that the stars are composed mainly of hydrogen prompted a reassessment of solar evolution and dynamics which, among other things, called for detailed elemental data.

Although it is targeted primarily at the photosphere, the results of spectroscopic study of the Sun are sometimes taken to apply to the Sun as a whole, but there has been some progress in aiming elemental analysis at different parts of the Sun and in identifying sources of distortion, to the benefit of our understanding of the Sun's internal dynamics. Even so the findings need to be supplemented, especially regarding elements which have no observable lines in the solar spectrum, by data from the solar wind, solar flares, sunspot spectra, solar energetic particles (SEPs), helioseismology, and theory[17]. In addition, CI chondritic meteorites provide complementary, independent checks on solar abundances on the assumption that they originated in asteroids in the primaeval nebula at the same time as the Sun and have not undergone significant or undetectable alteration.

Solar composition remains of critical importance for a wide range of investigations in astronomy—other than the purely stellar matters such as the refinement of Standard Solar Models (SSMs)—which range from planetary issues to the expanding redshift Universe[1]. Spectroscopy remains a key source albeit strongly dependent on advances in modelling of the solar atmosphere and of spectral line formation as well as in instrumentation and computing, and it benefits from almost a century of progress in atomic physics, solar modelling and observing technology, including devices which exploit artificial intelligence and are to some extent automated.

The Photosphere

The care taken with the precision of abundance measurements sometimes risks eclipsing any reservations about what it is that is being measured and whether sampling was judicious. Spectroscopic measurements portray the photosphere, which is considered to be sufficiently well mixed and shallow (~100 km) to avoid contamination with 'waste' from the nuclear core[35]; yet it is not unusual for the results to be labelled 'solar' and thought to reflect conditions at the birth of the solar system without taking into account aeons of chemical and gravitational (including rotational) distortion[2]. Hence the value of comparing the results with the composition of meteorites that are also thought to have retained the chemical signature of the parent nebula either unaltered or distorted in different ways.

When, in 1925, Cecilia Payne submitted her PhD thesis on *Stellar Atmospheres* to Radcliffe College, the subtitle read *A contribution to the observational study of high temperature in the reversing layers of stars*. The reversing layer is a term then used for the top of the photosphere. This was known to be the source of the dark Fraunhofer absorption lines that supplant the bright emission photospheric lines in the shortlived flash spectrum of a total eclipse (Figs. 3.1 and 3.2a)[22]. The flash spectrum in Fig. 3.2b shows the Fe XIV line, which corresponds with a temperature of 2 million K. The fragility of the visual data on which Payne and her colleagues depended is illustrated by the absence of the corresponding Fe XIV line, whose manifestation demands high solar activity and good weather during the peak or totality of the eclipse.

In her thesis Payne demonstrated that hydrogen was far more abundant in the stars than any other element. Demonstrated is not quite correct because she was dissuaded from explicitly saying so by one of her doctoral advisers. Payne had concluded that 'all the commoner elements found terrestrially …are actually observed in the stars' but that there is an 'enormous abundance of hydrogen and helium'[20]. Russell convinced Payne to tone down her revolutionary conclusions, and in the event she concluded that the abundance of hydrogen and helium was 'almost certainly not real', that the values she had calculated were 'regarded as spurious', and that hydrogen gave 'an impossibly high value'[20].

It has been argued[5] that Russell was acting in Payne's best interests given the novelty of the procedure by which she derived temperature from spectral evidence. In any event Russell later announced the 'almost incredible' abundance of hydro-

Fig. 3.1 Solar eclipse of
2017 showing chromosphere
(red) and corona (white)
(courtesy of NASA)

gen in the Sun's atmosphere, but, beyond stating that Payne's thesis was the 'most
important previous determination of the abundance of the elements by astrophysical
means'[26] and reporting agreement between their estimates, he did not make specific
acknowledgment of her priority in the key finding. The abundances determined by
Russell for 56 elements in the solar atmosphere remained the accepted standard for
many years.

Payne's thesis was an extraordinary achievement in three ways. First, for coun-
tering the established view of solar composition; second, for building on advances
in quantum physics and above all on the work by Meghnad Saha on the association between temperature and degree of ionization; and third, for demonstrating the
broad uniformity in stellar composition[36], which paved the way for work on the stellar origin of many elements by Fred Hoyle and others. In the absence of the internet
the work speaks highly for the observatory library and collection of stellar spectra
but above all for Payne's intellect, independence of mind, and alertness to crucial
advances in physics.

Payne focused on 18 elements and found that all but hydrogen and helium made
up in total a mere 2% of the matter in the visible stars. The crucial equation by Saha[27]
links the ionization state of an element, that is to say the extent to which its constituent
electrons have been jettisoned, with temperature and pressure. Saha showed that,
with increased temperature, stars with similar composition would display the spectral
types that had been classified by Antonia Maury and then Annie Cannon[31] as M, K, G,
F, A, B and O (Fig. 1.3c), and that, for stars at the same temperature, ionization would
increase with a decrease in pressure. Payne[20] was able to show that it was reasonable
to assume a maximum effective pressure of 1/10,000 atmospheres for the reversing
layer, thus eliminating one of the two variables in her assessment. The degree of
ionization, commonly expressed as the number of electrons lost, and conveyed by

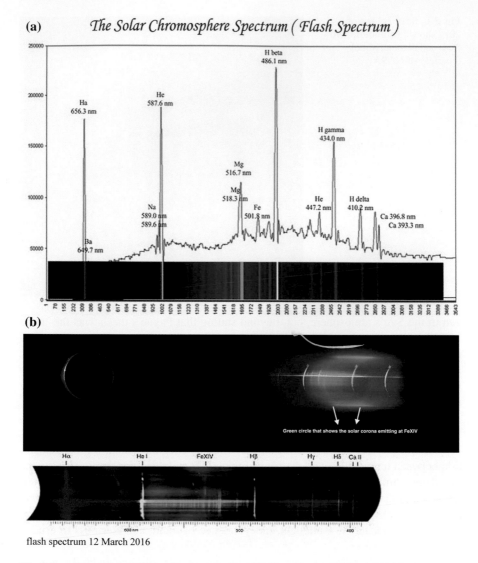

(a)   The Solar Chromosphere Spectrum (Flash Spectrum)

(b)

flash spectrum 12 March 2016

Fig. 3.2 a Composition of the solar chromosphere displayed in the 12 March 2016 flash spectrum (courtesy of NASA). **b** Flash spectrum for 12 March 2016. Fe XIV (green zone), evidence of a coronal temperature >2 million K, is generally displayed only when high solar activity coincides with excellent weather because the effect necessarily calls for short exposure times (image and interpretation courtesy of Constantinos Emmanoulidis)

roman numerals, could then be assessed by eye according to the 'estimates of width-intensity-contrast between the line and the continuous background' from glass plates (as shown by the example in Fig. 1.3b).

Modern stellar spectroscopy is of course no longer reliant mainly on the visual acuity and dedication of human observers. It employs photoelectric and electronic devices to record and quantify its subject matter and is now increasingly embracing automation[25]. Moreover various spectroscopic and other techniques encompass a far wider range of wavelengths than in the pioneering Harvard days. As it spans the radio and X-ray parts of the spectrum, as well as the visible wavelengths, astrospectroscopy requires satellite and rocket data for X-ray, UV and IR observation and antennas or dishes for the radio wavelengths.

Abundances

Absorption lines in the spectrum are expected to give abundance values with an accuracy of at best $\pm 10\%$, and they are in any case evaluated in the light of models of the solar atmosphere. The introduction of three-dimensional radiation hydrodynamic models and other refinements in about 2001 resulted in reductions of over 30% in the values for volatile elements such as C, N and O between estimates made in 2009 and those made in 1993 and 1997[30]. Abundances are not observed but inferred[17] from a combination of spectroscopic data with realistic models of the solar atmosphere and of the processes that give rise to the spectral evidence[2].

Photospheric composition is assumed by some to apply to the whole convection zone[2] and by others only to its outer part[14] but, as noted earlier, it is often extended to the Sun in its entirety ('solar abundances') on the grounds that, outside the core, mixing is thorough. This habit overlooks projects specifically designed to record the entire solar spectrum, or at any rate a substantial part of it, in order to portray the 'sun-as-a-star' for studies of long-term variations in irradiance, rotation or other unitary solar features[19, 21].

More important, it disregards the potential value of changes in composition over time and at different depths for our understanding of solar dynamics and for a correct assessment of abundances in the nascent Sun. And in any event there are no observable lines for ten elements—arsenic, selenium, bromine, tellurium, iodine, cesium, tantalum, rhenium, mercury, and bismuth—in the photospheric spectrum, and noble gas abundance have to be derived from other sources or theoretically[17].

The major processes at issue are element settling, mixing driven by the Sun's rotation, and chemical effects. That such processes were at work was suspected by Eddington[7] but they did not appear to explain abundances observed in other stars and in any case were not open to investigation until their calculated impact on sound velocity could be compared with the findings of helioseismology[33]. The outcome of these distortions may be expressed as a metal/hydrogen ratio, Z/X, where Z denotes the total mass fraction of heavy elements (quaintly termed metals and including oxygen, carbon, neon and iron) and X denotes the light elements hydrogen and helium. The present X/Y ratio (~ 0.0343) enters[13] critically into SSMs and a change in a single elemental measurement can have disproportionate implications here. Thus a reduction by a half of the assessed oxygen content of the solar atmosphere for some models points to a shallower convective zone.

Fig. 3.3 View of the Solar Wind Composition Experiment, a sheet of ultra pure aluminium and platinum metal foil, on the Moon. It was deployed during the Apollo 11, 12 and 14–16 missions to sample the solar wind outside the Earth's magnetosphere (courtesy of NASA)

Settling is driven by gravity and thermal diffusion, that is to say the separation of a mixture of gases where there is a temperature gradient, and it colours the interpretation of spectroscopy. Some of the models that underpin helioseismology initially assumed that the relative abundance of heavy elements at the core had remained unchanged, and they now benefit from theoretical approaches to the changing distribution of helium and heavy elements[32]. Indeed, helioseismology is sensitive enough to detect the impact of settling on a level of helium depletion in the convection zone amounting to a mere 8%. Note that solar rotation may have induced turbulence and therefore inhibited settling, and that nuclear processes below the convection zone destroy lithium but affect heavy elements only insofar as C and O are partly converted to N in the CNO cycle[11, 33].

Side by side with these calculations we may wish to sample parts of the solar atmosphere directly rather than following the indirect routes of spectroscopy or the mineralogy of meteorites, but we then need to allow for the first ionisation potential (FIP) effect, which depends on the amount of energy required to remove the outermost

(a)

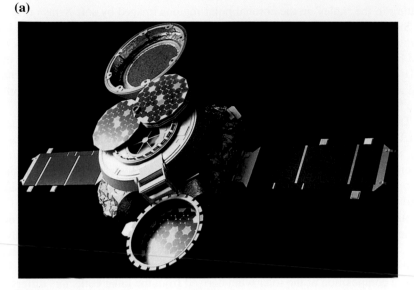

(b)

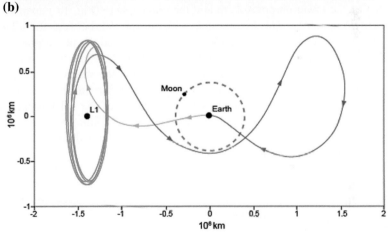

Fig. 3.4 a The Genesis spacecraft, which was targeted at the Sun's composition, collected solar wind particles in the course of 850 days. Besides high speed and low speed solar wind samples it returned coronal mass ejection material (courtesy of NASA). **b** Trajectory of the Genesis spacecraft showing the 5 halo loops it completed around the Earth-Sun Lagrange point L1

electron from a neutral atom in a gas (i.e. to ionise it) and which varies from element to element. For the core we may use the neutrino flux, as discussed in Chap. 5.

A tempting route to coronal analysis is to sample the solar wind, a plasma composed mainly of ionized hydrogen (electrons and protons) with a small percentage of helium (as alpha particles) and trace amounts of heavy ions and nuclei. It can be viewed as the outer corona in a state of steady expansion[22] which reaches at least to the heliopause where it meets interstellar space. Two categories of solar wind are recognised by their average speeds: fast, low density flow (>500 km/s), largely originating in the 'holes' in the coronal polar regions imaged by soft X-rays (Fig. 1.8) by Skylab astronauts, and slow, high density flows (<450 km/s) originating at lower latitudes. Field lines of the coronal magnetic field are embedded in the solar wind and help to define the heliosphere (Chap. 6). The corona and the solar wind in the polar regions are affected by the Sun's activity far less than those at lower latitudes and are therefore considered to be more representative of mass emission from the Sun[30].

The solar wind has been investigated in numerous ways. SOHO, which is parked at L1 and monitors the wind uninterruptedly, detected for the first time phosphorus, titanium and chromium as well as isotopes of iron and nickel not previously recorded[9]. Direct sampling has been done by way of metal foil targets left on the Moon by Apollo astronauts (Fig. 3.3) and by collectors on the Genesis mission in 2001–2004 (Fig. 3.4). A solar wind monitor has operated on the Advanced Composition Explorer (ACE) spacecraft close to the ecliptic—that is, the plane of the planetary o since 1997, and its measurements are complemented by those of Ulysses (Fig. 3.5) over the poles of the Sun[3] so that both slow and fast solar wind streams have been sampled.

The composition of the solar wind derived from the lunar collectors differs little from that of the lower chromosphere and the photosphere (Fig. 1.6) apart from its lower content in helium. This is consistent with the parallels between the solar wind speed and irradiance in the corona, the transition zone and the photosphere, which betray an intimate association between the solar atmosphere and the sunspot cycle, and thus with the solar convection zone[34]. To be sure, the various sources do not always agree. Part of the differences arise simply from the mode of measurement and the method of analysis, as well as the variability of the solar wind and the processes at work in the corona and chromosphere contributing to it. Again, with the exception of Ulysses the missions were confined to the ecliptic plane and their findings were thus dominated by the slow solar wind.

SEPs are high energy particles associated with CMEs—the 'gradual' category—and with episodes of magnetic reconnection—'impulsive' SEPs—and it has long been suspected that the composition of gradual SEPs is similar to that of

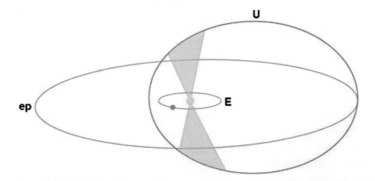

Fig. 3.5 Orbit of the Ulysses spacecraft (1990–2009), which was designed to investigate the heliosphere as a function of solar latitude and was the first to examine the Sun's poles. *U* Ulysses's orbit, with north and south polar passes shaded, *E* Earth's orbit around the Sun, *ep* Jupiter's orbit showing Ulysses' aphelion (right) in June 2004. Among the mission's targets were the Sun's magnetic field and the solar wind (courtesy of ESA)

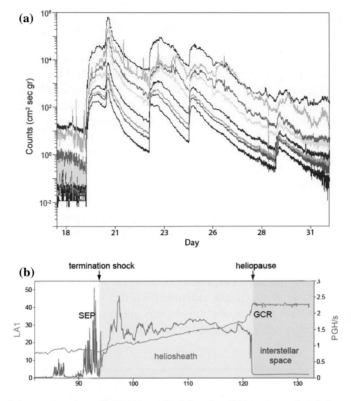

Fig. 3.6 **a** Solar proton events (SPE; >10 meV) of October 1989, compared with the proton storm of November 1997, recorded by GOES spacecraft at geosynchronous orbit. They peaked on 19–20 October. **b** Termination shock, heliosheath and heliopause encountered by Voyager 1 (*Credit* Stauriko under Creative Commons based on gsfc.nasa data). x axis in astronomical units; red line: SEPs, blue line: GCRs

the corona (Fig. 3.6). There is a problem with particle scattering en route to the observer, as shown by differences in the readings of the same event in the Wind and Ulysses spacecraft, but averaging of such variations provides credible guides to coronal abundances[24]. Additional motivation for their study should come from the evidence that high-energy SEPs and specifically protons with energies above about 50 meV pose a serious radiation hazard, especially outside the Earth's magnetic field, to astronauts and passengers on high altitude, transpolar routes, yet their onset and duration remains unpredictable[23].

Inevitably the favoured approach in solar modelling is to combine spectroscopic determination of volatile elements with meteoritic data for refractory elements. The meteorites of choice are the CI group of the carbonaceous chondrites, which have not been heated to melting temperatures as would be indicated by the presence of chondrules. They lend themselves to analysis by a range of techniques which yield very high precisions.

Meteorites

Meteorites, like the asteroids where most of them originated, are not simply condensed and unaltered versions of the primordial cloud but complex mixtures affected
by reaction with water and heat as well as the loss of volatiles and gases[12]. The CI
chondrites, which are represented by a mere six falls, are considered to be among the
most primitive meteorites so far investigated. Their composition is very close to that
of the photosphere (Fig. 3.7), and most differences can be explained by processes
acting in space, within the photosphere, or inside the asteroid that fathered the meteorite. For example, boron and lithium are destroyed in the nuclear fusion processes in
the Sun and are consequently depleted in the Sun relative to the chondrites whereas
carbon, nitrogen and oxygen are produced in nuclear fusion reactions in stars and are
therefore enhanced in the Sun relative to the chondrites. Naturally there was some
variation in the composition of the parent asteroids according to their location in
the nebula—some 27 asteroid types have been sampled[4]—but the great analytical
precision that can now be attained by mineralogists is thought to compensate for the
indirect character of the chondritic contribution to the analysis of the Sun as well as
for the small number of specimens available for direct study.

In the present context there are two obvious applications for elemental analysis
of the photosphere. First, as noted above, to exploit any divergence in composition

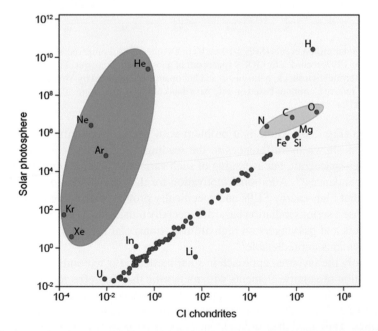

Fig. 3.7 Comparison between the composition of the solar photosphere and of CI chondritic meteorites. Note higher content in the chondrites of the volatile elements hydrogen, carbon, nitrogen
and oxygen and the noble gases. Lithium is depleted by nuclear processes within the Sun. (Based
on the Inspire website)

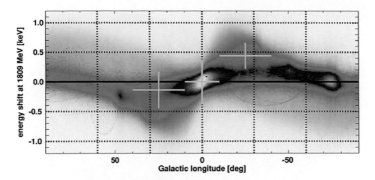

Fig. 3.8 Gamma rays produced in the galaxy by aluminium-26 and mapped using Doppler shift by ESA's Integral satellite. [26]Al is produced by supernovae and the gamma ray survey agrees with an estimated frequency of supernovae in our galaxy of 1 per 50 year (image courtesy of the Max Planck Institute for Extraterrestrial Physics, Garching). The galaxy's rotation causes blueshifts for negative galactic longitudes and redshifts for positive longitudes

from that of the primordial nebula, whether modelled or inferred from meteoritic data, in order to elucidate processes at work in the Sun's convection zone. Second, to extend our findings to the solar system as a whole and to elemental origins in general including generation in a supernova. The solar atmosphere is in a constant state of agitation, and the composition observed by spectroscopists reflects 4.56 Gyr of modification which may among other things result in partial ionization[11] and thus sabotage a diagnostic device that served Cecilia Payne well.

The last point might seem irrelevant to the study of the present-day Sun even if a supernova—Cassiopeia A—was recorded by John Flamsteed as recently as 1680, and type Ib/c and type II supernovae erupt in the Milky Way galaxy with a frequency of about one per 50 year (or more precisely 1.9 ± 1.1 per century), bringing them within the reach of our elastic definition of 'now'. Much stellar nucleosynthesis is of course associated with stars that are progenitors of Type II supernovae[8]. The frequency estimate is based on the measurement by ESA's Integral satellite of gamma rays produced by aluminium 26 ([26]Al; half life $t_{1/2} \sim 720{,}000$ year) (Fig. 3.8), a measurement which supersedes earlier values based on the rate in other galaxies or on supernova remnants in the Milky Way[6] and which indicates a current process rather than a residual effect of Solar System formation.

There remains the matter of sampling quality. CI chondrites are rare, fragile and inhomogeneous (Fig. 3.9). Powdered bulk samples of the Orgueil meteorite contain magnetite, dolomite and phyllosilicates yet X-ray diffraction failed to detect dolomite in a sample even though electron microscopy in other parts of the fragment revealed 5% of dolomite (calcium magnesium carbonate), an important guide to the degree of alteration[15]. Mineralogists have long agonised over the procedures required to ensure that a sample is representative of a population, or an environmental condition, and that it embodies both accuracy and precision[18], that is the 'truth' of the analysis as well as the reproducibility of the result.

Fig. 3.9 Part of the Allende carbonaceous chondrite. Note the fusion crust and the white calcium-aluminium-rich inclusions (CAI). The specimen is about 10 cm long (image by the Natural History Museum London)

A simple ruse to achieve this is termed modal analysis where a square grid is superimposed on a thin microscope section and the mineral species at the intersections of the grid are recorded[16], but the nature of the grid and the sampling routine remain arbitrary. Random sampling is equally elusive and pure randomness may lead to sampling which is not representative. Even so sampling procedure as well as analytical methods influence the results of meteoritic profiling so subtly that the close resemblance between the photospheric and meteoritic readings, with a few exceptions, is surprising, and one is led to suspect that the reported level of analytical precision is too gross to identify the impact of solar history on present day abundances and their meteoritic counterparts. Refinement by one order of magnitude, probably already lurking in laboratory archives, could suffice to reveal differences. Fortunately helioseismology provides another way of uncovering processes at work in the solar subsurface (see Chap. 4).

Besides parochial issues relating to the processes at work in the Sun there is much to be learnt about galactic evolution from elemental study. It is true that we understand the evolution of baryonic matter during the first three minutes of primordial nucleosynthesis better than for the last 5 Gyr in the life of the galaxy[10] but only if the questions are framed in the language of electron pairs, photons and neutrinos. The chemical evolution of the 'highly heterogeneous, evolved universe', with or without the added ingredient of dark matter, is indeed more elusive, but, as often, a reductionist program can yield unexpectedly far-reaching results: in the present case, by illustrating processes at work at a mid-point in the evolution of a main-sequence star.

References

1. Asplund M, Grevesse N, Sauval AJ (2005) The solar chemical composition. In Barnes TG III & Bash FN (eds) Cosmic abundances as records of stellar evolution and nucleosynthesis, ASP Conf 336:25–38
2. Asplund M et al (2009) The chemical composition of the Sun. Annu Rev Astron Astrophys 47, 481–522
3. Balogh A, Marsdenet RG, Smith EJ (2001) The heliosphere near solar minimum: the Ulysses perspective. Springer, Berlin
4. Campbell IH, O'Neill HSC (2012) Evidence against a chondritic Earth. Nature 483:554–558
5. DeVorkin DH (2010) Extraordinary claims require extraordinary evidence: CH Payne, HN Russell and standards of evidence in early quantitative stellar spectroscopy. J Astron Hist Heritage 13:139–144
6. Diehl R et al (2016) Radioactive ^{26}Al from massive stars in the Galaxy. Nature 439:45–47
7. Eddington AS (1926) The internal constitution of the stars. Cambridge Univ Press, Cambridge
8. François P et al (2004) The evolution of the Milky Way from its earliest phases: constraints on stellar nucleosynthesis. Astron Astrophys 421: 613–621
9. Galvin AB and 26 others (1996) Solar wind composition: first results from SOHO and future expectations. Bull Am Astr Soc 28: 897
10. Geiss J, Gloeckler G (2007) Linking primordial to solar and galactic composition. Space Sci Rev 130: 5–26
11. Gorshkov A B & Baturin V A 2008 Diffusion settling of heavy elements in the solar interior. Astron Rep 52:760–771
12. Gounelle M. and Zolensky M. E. 2014. The Orgueil meteorite: 150 years of history. Meteoritics and Planetary Science 49:1769–1794
13. Grevesse N, Noels A (1993) Origin and evolution of the elements. Cambridge Univ Press, Cambridge
14. Grevesse N & Sauval AJ (2002) The composition of the solar photosphere. Adv Space Res 30:3–11
15. King AJ et al (2015) Modal mineralogy of CI and CI-like chondrites by X-ray diffraction. Geochim Cosmochim Acta 165, 148–160
16. Koch GS, Link RF (1970) Statistical analysis of geological data. Wiley, New York
17. Lodders K (2003) Solar system abundances and condensation temperatures of the elements. Astrophys J 591: 1220–124
18. Myers JC (1997) Geostatistical error management. Van Nostrand Reinhold, New York
19. Neckel H (1994) Solar absolute reference spectrum. Int Asr Un Colloq 143:37–44
20. Payne CH (1925) Stellar atmospheres. Harvard Univ Press, Cambridge Mass
21. Pevtsov AA, Bertello L, Marble AR (2004) The sun-as-a-star solar spectrum. Astron Nachr 335: 21–26
22. Phillips KJH (1992) Guide to the Sun. Cambridge Univ Press, Cambridge
23. Reames DV (2013) The two sources of solar energetic particles. Space Sci Rev 175:53
24. Reames DV (2014) Element abundances in solar energetic particles and the solar corona. Solar Phys 289:977–993
25. Rosenberg DE (2016) Automation of spectroscopic observations on the Dark Sky Observatory 32-inch telescope. MSc thesis, Appalachian State University
26. Russell HN (1929) On the composition of the sun's atmosphere. Astrophys J 70: 11–82
27. Saha MN (1920) Ionization in the solar chromosphere. Phil Mag 40:472–488
28. Schmelz JT et al (2012) Composition of the solar corona, solar wind, and solar energetic particles. Astron J 755:33
29. St. John CE, Babcock HD (1924) Pressure and circulation in the reversing-layer of the sun's atmosphere. Astrophys J 60:32–42
30. Serenelli A (2016) Alive and well: a short review about standard solar models. ArXiv:1601.07179 v1[astro-ph.SR]

31. Sobel D (2016) The glass universe. Fourth Estate, London
32. Turcotte S, Christensen-Dalsgaard J (1998) The effect of differential settling and the revised abundances on solar oscillation frequencies. ESA SP-418, Boston, 561–565
33. Vauclair S (2003) Diffusion and mixing inl main-sequence stars. Astrophys Space Sci 284:205–215
34. Vita-Finzi C (2016) The contribution of the Joule-Thomson effect to solar coronal heating. ArXiv:1612.07943
35. Von Steiger R et al (2001) Measuring solar abundances. In Wimmer-Schweingruber RF (ed) Solar and galactic composition. AIP Conf Proc 598, Melville, NY, USA, 13–22
36. Wayman PA (2002) Cecilia Payne-Gaposhkin: astronomer extraordinaire. Astron Geophys 43:1.27–1.29

Chapter 4
The Solar Body

Abstract In the 1960s localised oscillations of the Sun's surface were found to represent acoustic and other internal waves which were to reveal many features and processes of the solar interior. In contrast, despite prolonged observation of the Sun from telescopes, balloons and orbiting satellites, there is uncertainty over any surface evidence for gravitational interplay with the planets, the Schwabe and other activity cycles, and the mass loss expected from the solar wind and nuclear fusion. But the variegated ramifications of the subject, including relativity and stellar evolution, guarantee further astrometric innovations and refinements.

Helioseismology was born a mere 8 years after oscillations of the Sun's photosphere were accepted as real[18, 33]. In contrast, although the study of solar shape and dimensions stretches back to Greek antiquity[28], and there have been great advances in the relevant instrumentation and in our understanding of solar physics, the outcomes continue to provoke controversy. It may be that measuring a ball of hot gas is destined to be problematic, even without the obstruction of a turbulent atmosphere around the observer, but fortunately the wider implications of the subject, including tests of the predictions of general relativity, spur on observation and analytical innovation.

The dedicated space probes that have investigated the Sun's nature and behaviour include one named after Jean Picard, the astronomer who performed the classic measurements of solar diameter in 1666–1719. For four years the spacecraft targeted the Sun's volume and shape, its differential rotation at different wavelengths, total and spectral solar irradiance, the Sun's role in creating and destroying the Earth's ozone blanket, the transit of Mercury, and a number of observations of value to helioseismology. In addition, the solar factor in weather, as well as the measurement of solar diameter, were pursued by concurrent measurements from the ground.

Running the various enquiries more or less simultaneously gamely acknowledged that none of them would wholly succeed in isolation.

Solar Volume

Though commonly paraded for awesome comparison with the Earth (Fig. 4.1), the Sun's volume (1.4×10^{27} m^3) enters surprisingly little into any discussion of the dynamics of the solar system, perhaps because of uncertainty over the precise value of the gravitational constant. But there are two topics on which the solar volume per se cannot be sidestepped: cumulative mass loss and the orbital implications of relativity.

The fusion process at the Sun's core leads to a reduction in hydrogen relative to helium, with the generation of radiation and neutrinos, at a rate of about 4.4×10^9 kg/s. The second source of mass loss is occasioned by the solar wind, which conveys electrons and protons to space at a rate of about 1.4×10^9 kg/s or 10^{-13} $M_\odot$/year[22]. A third, less easily quantified, component is the ejection of matter in solar storms, particularly CMEs[30], which have been called an extreme version of the solar wind as

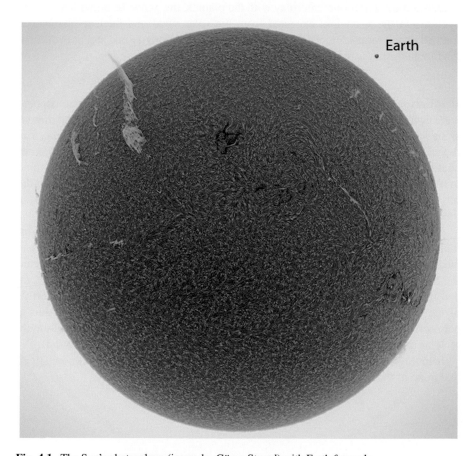

Fig. 4.1 The Sun's photosphere (image by Göran Strand) with Earth for scale

they contain high speed plasma and magnetic field[11]. Their mass loss contribution is about 10–15% of that due to the background solar wind[19], say 10^8 kg/s. As a check we may combine the widely cited average mass loss per CME of 10^{12} kg with the largest annual number of CMEs catalogued from SOHO by the year 2004, namely 1652 for the year 2002[36] to obtain a total mass loss which is again ~10^8 kg/s.

A 'constant rate' for H–He conversion is evidently an approximation as the process hinges among things on the Sun's age and thus a progressive decrease in the H:He ratio. Moreover the dominant process is the proton-proton chain reaction (Fig. 5.1) with components that operate at widely varying rates even if these variations are broadly ironed out in the long-term. In the short term, however, fluctuations in the solar neutrino flux (Fig. 5.2) that are sometimes dismissed as mere casualties of measurement may reflect substantial oscillations in the p-p process and in the Sun's internal dynamics[29].

Solar mass loss has many implications even if most of them are likely to elude direct observation for some time to come. They include the dynamic definition of AU, which (as in this book) is generally defined as the average of the distance between Earth and Sun at aphelion and perihelion, and is widely used as yardstick for distances in the solar system as well as contributing to the definition of a parsec (Fig. 4.2). The alternative definition of AU relies on Johannes Kepler's Third Law, which states that the squares of the sidereal periods (P) of the planets are proportional to the cubes of the semi-major axes (a) of their orbits (Fig. 4.3).

The mass loss would lead to expansion of planetary orbits and a reduction in their angular motion, as the Sun's gravitational grip on them is loosened. The corresponding increase in the AU is perhaps 2 cm/year[22, 28], although radar ranging of the inner planets and of a number of orbiters and landers, notably the MESSENGER spacecraft, suggests it will amount to as much as 15 cm/year[13]. Any disparity between the measured and the estimated rates of increase could conceivably result from a change in solar wind flux, which would obviously affect the mass loss figure, or from clock drift, which might distort spacecraft data[17]. But the agreement between estimate and measurement is compelling enough.

Mercury's ephemerides (coordinates) gathered by MESSENGER over 7 years have supplied one of the first experimental checks on the mass loss estimates. They reveal a change in the Sun's gravitational parameter (i.e. $M_\odot$ × the gravitational constant G) which, after allowing for the timing of the mission in relation to the solar cycle, was consistent with a mean annual solar mass loss of ~$10^{-13} M_\odot$[8], in agreement with our earlier estimate.

Another possible though smaller influence on the AU is obtained by invoking the kind of tidal mechanism familiar to us from the Moon-Earth system: the Earth's mass creates a tidal bulge in the Sun, which slows it by 3 ms/100 years, and the reduced angular momentum is transferred to the Earth leading to an enlargement of its mean orbit. The change in the Sun's period of rotation, it has bizarrely been claimed, is too small to be ruled out by observation[20].

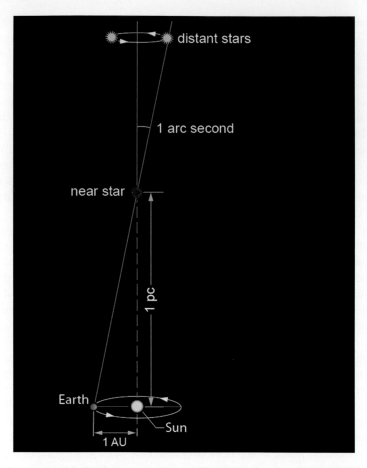

Fig. 4.2 A parsec (pc) is the distance at which 1 AU subtends 1 arcsecond (1/3600 of a degree), and is roughly equivalent to 3.26 light years (ly). The nearest star, Proxima Centauri, is ~1.3 parsecs from the Sun

    Throughout these tribulations the solar mass is implicitly treated as a point gravitational value when some 'asphericity' is generally accepted and equated with oblateness or polar flattening. The assumption here is that the driving forces are dynamic and that the response is axisymmetrical while internal processes can be neglected. Nevertheless analysis of data obtained by the Michelson Doppler Imager (MDI) on SOHO in 1997, which was subject to exhaustive procedure for image stabilization, revealed anomalous areas of surface brightness amounting to 1.5 K. They were not attributable to magnetic fields in the photosphere and prompted the suspicion that they might 'shadow' cyclical changes in the Sun's interior. Moreover analysis over two months at 10 mas sensitivity yielded a shape term which showed a 'marginally

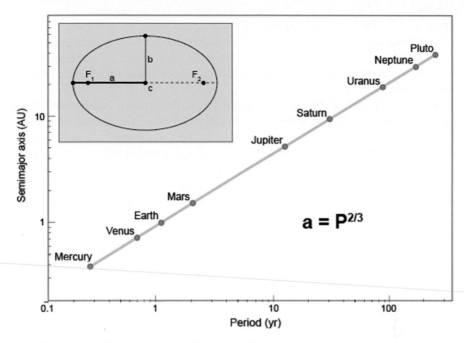

Fig. 4.3 Kepler's Third orbital law, published in 1619, states that the square of the orbital period of a planet (in Earth years) is proportional to the cube of the semi-major axis of its orbit (in AU). The inset shows the semi-major axis and the two foci of the elliptical orbit F1 and F2

significant' variation with the solar cycle[14, 15]. The hunt for periodicities may seem driven but any such link would provide a valuable supplement to probing by neutrino analysis.

Solar Diameter

It was perhaps inevitable that an equatorial bulge should be the item that has most exercised students of the solar body: both naked eye and telescopic observation yield an essentially two-dimensional image, and oblateness entered into solar-system thinking once Galileo mistakenly identified rotational flattening on Saturn. Observational history accounts for the reliance on angular measure (and by implication on an agreed length for the AU used in the trigonometric calculation) for much of this work.

Monitoring the Sun from outside the Earth's atmosphere did not resolve the issue, and uncertainty remains over two matters: the extent of any flattening and whether it is permanent, periodic or transitory. The Earth and the other terrestrial planets permit altimetry over land and water by orbiting satellites using lasers, radar and gravimetry[34]. The orbits of Ulysses, the sole space probe to orbit over the Sun's poles, did so at about 1.34 AU at perihelion and at about 5.4 AU at aphelion, so that variations from a modelled shape were not in its programme. Assessment of any departure from sphericity has therefore generally been based on measurements of the solar radius ($R_\odot$).

There are two major approaches to solar radius: photospheric and seismic[9]. They yield measures which may differ by as much as 0.3 Mm. The former, as we saw, dates from the dawn of astronomy. The techniques that have been employed to measure the photospheric solar radius include visual observation based on meridian timings, the micrometer and the astrolabe (reborn as the Danjon astrolabe); automatic measurement by photoelectric devices; and planetary transits and eclipses[24, 35]. The angular measurements they have provided since 1715, equivalent to a linear radius of 695–970 km, indicate less than ±2 arcsec deviation from the mean[23] with an estimated error of 20 mas (14.5 km).

The results obtained by different photospheric groups can differ by ~500 km thanks to systematic errors from different instruments and observers, statistical procedures, and varying definitions of the photosphere itself[6]. Thus some photospheric definitions use as solar edge the inflection point of the limb darkening effect on the Sun, which arises because the line of sight into the solar interior is oblique near the limb and therefore encounters opacity at a shallower depth than near the Sun's centre (Fig. 4.4); the method gives a different result according to the wavelengths employed in the observation[10] and its interpretation cannot ignore the possibility that rotation varies with depth as well as latitude[26].

Fig. 4.4 Limb darkening occurs because only the upper cooler layers of the solar atmosphere are seen at the limb and moreover are viewed obliquely. The effect is used to investigate the temperature structure of other stars

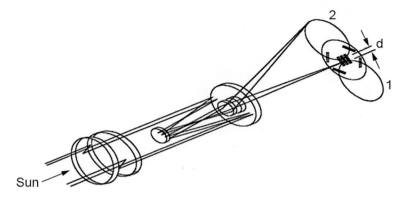

Fig. 4.5 In the Solar Disk Sextant, a beam-splitting wedge produces a direct image and an offfset reflected image (1, 2), and the diameter of the solar limb is determined by measuring the discs' separation and the gap d (after Sofia et al.[32])

Nor does evading the atmosphere simplify matters, sometimes for homely reasons. The balloon-borne Solar Disk Sextant (SDS: Fig. 4.5)[32] employed a beam-splitting wedge which provided dual images of the Sun's half-diameter. The system achieved the stability of 12 arcsec or better over the 10–12 h flights, which were confined to the times when high-velocity stratospheric winds change direction and the Sun is not too high in the sky, namely September 25 or later. Deployed seven times between 1992 and 2011 and targeted on 565–665 nm, it found a variation of up to 200 m.

The seismic approach has a much shorter history than the photospheric, as it dates from the 1970s. Data for surface gravity (f-modes) gathered from SOHO and the SDO over almost two solar cycles (1996–2017) appear to indicate variations in solar radius that are inversely related to sunspot number[27], whereas oscillations recorded by the MDI instrument in 1996–1998 have been dismissed as the result of temperature changes within the spacecraft[5]. Note that a later set of SDO measurements (2010-2012.5), this time using the Helioseismic and Magnetic Imager (HMI), which has 16 times as many detector pixels as in the MDI, indicated an oblate shape almost immune from solar-cycle variations on the Sun's surface apart from the hint of a correlation with sunspot number[16].

The difficulty of securing consensus is compounded when the subject is both mobile and blurred. It follows that narrowing the target in space and time can be rewarding, if at the cost of shelving long-term changes. For, example, as the flux of energetic photons from the Sun's core is in effect constant when viewed at the solar cycle timescale, there must be an 'intermediate energy reservoir' between it and the photosphere, and its nature should be reflected in changes in solar diameter and contemporaneous fluctuations in luminosity[5]. In order to discriminate between alternative models for the association between the two measures, as luminosity variations over the solar cycle averages only 0.1%, detection of the corresponding change in solar radius will need to be 70 mas or better[5, 31].

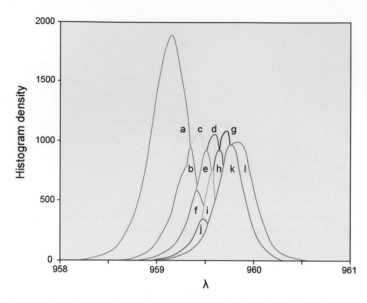

Fig. 4.6 Density plot of mean solar radius in as determined by the LATMOS instrument of the Picard ground mission (adapted from Ikhlef et al.[12]). Note dominance of 959.2 nm in the radius measurements

Again, it clearly pays to specify the wavelength at which the measurement is being made (Fig. 4.6) as it reflects the height of line formation. For example a measurement taken in the continuum wing of the 617.3 nm line resulted in a solar radius (corrected to 1 AU) of 959.57″ whereas the AIA instrument observing a transit of Venus at ultraviolet wavelengths[6] gave values of 963.04″ at 160 nm and 961.76″ at 170 nm.

It is of course no longer tenable to assume that deformation is confined to oblateness or dependent on some continuous (e.g. power law) coreward increase in density. Thus MDI data for 1997 and 2001 showed a decrease in equatorial oblateness combined with a mid-latitude depression[26], and in any case the picture will be complicated if rotation is differential with latitude as well as depth and on an axis which is not normal to the ecliptic plane. The Sun's oblateness is small (1/10,000 compared to the Earth's 1/298) but not simple and incompletely understood notably as regards velocity variations in the interior and distortion by magnetic factors[25]. That is not all. At times of high solar activity the Sun displays a pattern of 'cantalaoupe ridges' which border supergranules; this increases the equatorial radius compared to the polar radius[21] by 6 km over a total of 695,700 km or 0.0009%.

This brings us to the wider significance of solar deformation, notably the Sun's planetary interaction with other solar system bodies. The two-body gravitational problem, say that of Sun-Earth, is comfortably resolved in newtonian terms. The n-body problem would be troublesome to solve if the planets were not so far apart and dwarfed by the mass of the Sun, as their orbits can be resolved one at a time by the two-body procedure 'perturbed' by other planets. This is the route that led Le

Fig. 4.7 The precession of
Mercury's orbit around the
Sun should amount to 5557
arcsec/century. Calculated on
Newtonian principles it left a
deficit of 43 arcsec/century;
the gap was closed by
appealing to General
Relativity

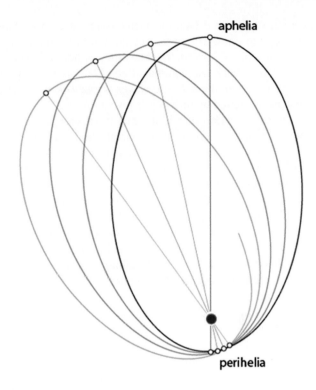

Verrier and Adams in 1846 to the discovery of Neptune, as the known planets could
not account in its entirety for the observed perturbation of Uranus.

For the Sun-Mercury two-body system the solar component would contribute
5030 arcsec/century for precession of the perihelion of Mercury, whereby the long
axis of the elliptical orbit rotates around the Sun by 5600 arcsec/century in response to
the gravitational effect of the oblate Sun (Fig. 4.7). An additional 530 arcsec/century
could be ascribed to the influence of other planets on Mercury's orbit. That left about
43 arcsec unexplained. General relativity was to provide the missing values.

Waves and Vibrations

In 1960–1961 'bright elements' of the solar surface measuring ~1700–3500 km
according to the wavelength employed were cautiously described[18] as 'quasi-
oscillatory' with a period of 296 ± 3 s (that is ~5 min), a vertical velocity averaging
0.15 m/s and a mean life of ~380 s. They occurred side by side with the cellular
pattern of supergranulation (see Chap. 5). The cycle was repeated about 6 times
before the process was initiated elsewhere on the solar surface. The background
noise against which the oscillations were measured was 330 m/s^2—the bounce of a

raft in a typhoon-driven sea. Detection hinged on the Doppler effect on a number of absorption lines. The initial interpretation favoured local deformation perhaps due to convection driven by 'the buoyancy of hot granules'[18]; eight years later came the realization that it was a global effect. The 5-min oscillations were now viewed as the impact of standing acoustic waves trapped beneath the photosphere, waves which had been excited by turbulent convection[7, 33].

It was then also surmised[33] that if wavelength was plotted against frequency the waves would be confined to discrete lines or ridges, as they are standing waves responding to different acoustic cavities inside the Sun. This is just as well from our point of view as there are some 10 million resonant sound (p or pressure mode) waves milling about in the solar interior[2]. By 1975 observations confirmed the standing-wave model, and the analysis yielded 3 or 4 stable modes for the 5-min oscillation[3].

The Sun also experiences g modes, which are standing gravity waves, but they are confined to the atmosphere or to the solar interior below the convection zone, and they are difficult to observe at the solar surface as they have amplitudes <3 mm, too small for the GOLF (Global Oscillations at Low Frequencies) system. On the other hand a further category, the f modes, which are surface gravity waves, are of great value in analysing such dynamic minutiae as the effect of Coriolis forces—that is the deflection in the path of a body or mass over a rotating sphere—or the flow of supergranules on the solar surface[4].

Unlike Doppler measurements, f mode data are not limited to line-of-sight observation. But the sensitivity now attained by the Doppler technique permits photospheric distortion amounting to a mere 1/100,000 of the solar radius, to be identified by networks of ground observatories as well as satellites. In particular, full-disk observation by GONG (the Global Oscillation Network Group), consisting of six ground observatories, and the MDI on SOHO supplanted in 2012 by the HMI instrument on the SDO, underpin the blossoming field of helioseismology. 'Seismology' is a misnomer insofar as, unlike the terrestrial version, we are dealing with gas rather than rock, and no shear (side-to-side) waves can operate. But it is apt in the sense that we are probing subsurface structures using oscillations detected at the surface; the material properties traversed by the waves can to some extent be assessed; and, as in geoseismology, there are ramifications: in chronology and in dynamics.

The ramifications are well illustrated by the base of the convection zone where it meets the radiative zone. The boundary—the tachocline—evidently defines different sound speed regimes, given its basis in helioseismology. Analysis of p-modes places it at 0.711 $R_\odot$, that is 71% of the way from the core to the surface. The tachocline demarcates zones with contrasting rates of rotation (Fig. 4.8). The upper convective zone, like the photosphere, is characterised by faster rotation at high latitudes, so that at 75°N, for example, the rate amounts to 33.4 days for a full rotation whereas at the equator it is 25.7 days. These values are not wholly consistent over time. Nor are they maintained consistently with depth. In contrast the radiative zone is considered to undergo solid body rotation, but the rate is decidedly wobbly.

The sound speeds calculated for the BP 2000 version of the SSM show excellent agreement with those obtained by helioseismology[1]. Perhaps more important, helioseismology also bears on helium abundance, the former an essential feature of models

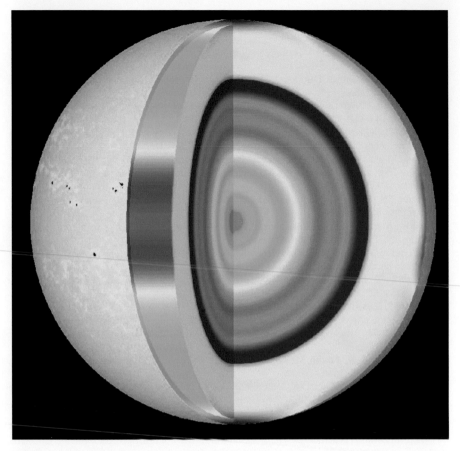

Fig. 4.8 Sound speed measured by instruments on SOHO point to a rapid change in rotation at the transition zone between the convection and radiative zones (prominent red band) and temperatures 0.1% lower than expected at the core (blue zone) (image based on ESA/NASA data by A Kosovichev, reproduced with permission)

of the Sun's interior, the latter a check on the hydrogen/helium ratio to be expected after the reactor has been switched on. Indeed, it has led to refinements in the solar composition as a whole and in particular the heavy element 'metal' component.

References

1. Bahcall JN, Pinsonneault MH, Basu S (2001) Solar Models: currently epoch and time dependences. Astrophys J 555:990–1012
2. Demarque P, Guenther DB (1999) Helioseismology: probing the interior of a star. Proc Nat Acad Sci 96:5356–5359
3. Deubner F-L (1975) Observations of low wavenumber nonradial eigenmodes of the Sun. Astron Astrophys 44:371–375
4. Duvall TL, Gizon L (2000) Time-distance helioseismology with f modes as a method for measurement of near-surface flows. Solar Phys 192:177–191
5. Emilio M et al (2000) On the constancy of the solar diameter. Astrophys J 543:1007–1010
6. Emilio M et al (2015) Measuring the solar radius from space during the 2012 Venus transit. Astrophys J 798:48
7. Frazier EN (1968) A spatio-temporal analysis of velocity fields in the solar photosphere. Z Astrophys 68:345–356
8. Genova A et al (2018) Solar system expansion and strong equivalence principle as seen by the MESSENGER mission. Nature Comm 9:289
9. Haberreiter M, Schmutz W, Kosovichev AG (2008) Solving the discrepancy between the seismic and photospheric solar radius. Astrophys J 675:L53-L56
10. Hestroffer D, Magnan C (1998) Wavelength dependency of the solar limb darkening. Astron Astrophys 333:338–342
11. Howard T (2014) Space weather and coronal mass ejection. Springer, Dordrecht
12. Ikhlef R et al (2013) PICARD-SOL. Presentation and results. CNES, Paris
13. Krasinsky GA, Blumberg VA (2004) Secular increase of astronomical unit. Dyn Astro 90: 267–288
14. Kuhn JR, Bush R, Emilio M, Scholl IF (2012) The precise solar shape and its variability. Science 337:1638–1640
15. Kuhn JR, Bush R, Scheick X, Scherrer P (1998) The Sun's shape and brightness. Nature 392:155–157
16. Kuhn JR et al (2012) The precise solar shape and its variability. Science 337:1638–1640
17. Lämmerzahl C, Preuss O, Dittus H (2017) Is the physics within the solar system really understood? In Dittus H et al (eds) Lasers, clocks and drag-free control, Springer Heidelberg, 75–101
18. Leighton RB, Noyes RW, Simon GW (1962) Velocity fields in the solar atmosphere. I. Preliminary report. Astrophys J 135: 474–499
19. Mishra W et al (2018) Solar cycle variation of coronal mass ejections contribution to the mass flux. arXiv:1805.07593v1[astro-ph.SR]
20. Miura T et al (2009) Secular increase of the Astronomical Unit: a possible explanation in terms of the total angular momentum conservation law. arXiv:0905.30008v3 [astr-ph.EP]
21. NASA (2008) Sun not a perfect sphere. Science News 6 October 2008 at sciencedaily.com
22. Noerdlinger P (2008) Solar mass loss, the Astronomical Unit, and the scale of the Solar System. arXiv:0801.3807 [astr-ph]
23. Parkinson JH, Morrion LV, Stephenson FR (1980) The constancy of the solar diameter over the past 250 years. Nature 288:548–551
24. Ribes E et al (1991) The variability of the solar diameter. In Sonett CP, Giampapa MS, Matthews MS (eds) The Sun in time, Univ of Arizona, Tucson, 59–97
25. Rozelot JP, Damiani C (2011) History of solar oblateness measurements and interpretation. Europ Phys J H 36:407–436
26. Rozelot JP, Fazel Z (2013) Revisiting the solar oblateness: is relevant astrophysics possible? Sol Phys 287:161–170
27. Rozelot JP, Kosovichev AG, Kilcik A (2018) How big is the Sun: solar diameter changes over time. Sun Geosph 13:63–68
28. Rozelot JP, Kosovichev AG, Kilcik A (2016) A brief history of the solar diameter measurements: the critical quality assessment of the existing data. arXiv:1609.02710

29. Sakurai K, Haubold HJ, Shirai T (2911) The variation of the solar neutrino fluxes over time in the Homestake, GALLEX(GNO) and Super-Kamiokande experiments. arXiv:1111.5530v1 [physics.gen-ph]
30. Sheeley NR et al (1999) Continuous tracking of coronal outflows: two kinds of coronal mass ejections. J Geophys Res: Space Phys 104:A11
31. Sofia S, Endal AS (1979) Solar constant: constraints on possible variations derived from solar diameter measurements. Science 204:1306–13084.
32. Sofia S et al (2013) Variation of the diameter of the Sun as measured by the Solar Disk Sextant (SDS). MNRAS 436:2151–2169
33. Ulrich RK (1970) The five-minute oscillations on the solar surface. Astrophys J 162:993
34. Vita-Finzi C (2002) Monitoring the Earth. Terra, Harpenden
35. Vita-Finzi C (2013) Solar history. Springer, Dordrecht
36. Yashiro S et al (2004) A catalog of white light coronal mass ejections observed by the SOHO spacecraft. J Geophys Res 109, A07105
37. Zuber M, Smith DE (2017) Measuring solar mass loss and internal structure from monitoring the orbits of the planets. Lunar Planet Sci LVII, #2281

Chapter 5
The Solar Furnace

Abstract The accepted scheme for the solar interior is a core powered by nuclear fusion primarily via the p-p reaction. The validity of the Standard Solar Model for the Sun's related metabolism can be assessed by reference to the associated flux of neutrinos. A temperature of 15 million K at the core falls to ~6000 K at the photosphere and rises to 20,000 K at the chromosphere and to 1–2 million K or more at the outer margins of the solar corona; most of the mechanisms currently favoured for this progression hinge on magnetic reconnection or on wave heating, whereas the scheme advanced here is a stepwise sequence, with induction heating at the photosphere, the Joule-Thomson effect in the chromosphere, and plasma expansion in the corona, a tripartite solution which also explains the threefold structure of the solar atmosphere.

The processes that have been advanced to account for the Sun's luminosity range from the red hot stone envisaged in ancient Greece, through the impacting meteors or gravitational contraction of mid-19th century science, to the modern notion of a nuclear core where energy is released mainly by the proton-proton (p-p) chain reaction[37] and eventually reaches the photosphere to be manifested as various forms of radiation.

Some of the energy generated by the p-p process is in the form of neutrinos, which reach the Earth at a rate calculated using solar luminosity of about 7.10^{10} per cm^2 per second. Figure 5.1 shows the neutrino flux on Earth for two different fusion reactions. Note the large flux (86% of the total) and low energy of the proton-proton reaction compared with the relative low flux (14%) and high energy of the ^8B route. Variations in this flux thus provide a direct probe into the core as well as the zones traversed by the escaping neutrinos, and spectral detection of p-p neutrinos has demonstrated that 99% of the Sun's power is generated by that process[11].

© Springer Nature Switzerland AG 2018
C. Vita-Finzi, *The Sun Today*, https://doi.org/10.1007/978-3-030-04079-6_5

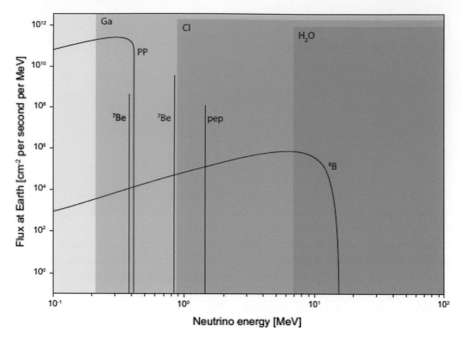

Fig. 5.1 Neutrino flux for two of the reactions predicted by the Standard Solar Model of 2005—the pp chain and ^{8}B (after Bahcall et al.)

Other tests of a Standard Solar Model include data obtained by helioseismology and more conventional measures of solar behaviour. How the corona attains temperatures of one or more million K remains in dispute. Most of the favoured explanations invoke magnetic reconnection or some kind of wave heating. The alternative proposed here is tripartite, with different mechanisms operating in the photosphere, chromosphere and corona.

The Nuclear Core

Legend has it that the current conception of the Sun's inner workings dates from a rail journey undertaken by the physicist Hans Bethe after a 1938 conference in Washington on stellar energy[13]. The legend has been trashed (although train journeys undertaken by Bethe in 1934 and in 1947 yielded important results) but Bethe's paper on energy production in stars did come out soon after the 1938 meeting[8, 11]. It established a process by which, in faint stars like our Sun, carbon and nitrogen act as catalysts in producing a ^{4}He nucleus from hydrogen atoms by way of the proton-proton or p-p chain (Fig. 5.2). The corresponding scheme for massive stars (i.e. those with 4 solar masses or more) is the carbon-nitrogen (CNO) cycle, now thought to contribute only 7% of the power yield by the Sun.

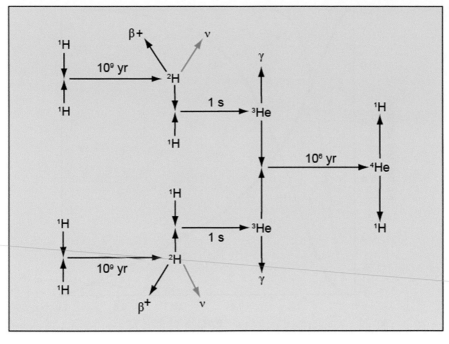

Fig. 5.2 The proton-proton (p-p) chain, simplified after Eric Blackman, with permission[8]. It dominates in stars with a mass of the size of the Sun or smaller. Some energy is contributed by neutrinos (ν) and gamma rays interacting with electrons and protons

The very high temperatures required for nuclear fusion to start can result from the gravitational energy released by contraction of the solar mass. In a word, the nineteenth century physicists who favoured gravitational contraction to heat the Sun have been allowed to contribute the trigger.

The current Standard Solar Model (Fig. 5.3) is devised by identifying the processes that make a symmetrical plasma sphere of one $M_\odot$ and of zero age (ZAMS: zero age main sequence i.e. when the Sun entered the Main Sequence on a Hertzsprung-Russell plot) eventually fit the observed luminosity, radius, age and composition of today's Sun. This is not unlike the procedure followed by a physician seeking to discover the explanation for the symptoms he is now observing in a patient. The 1998 SSM, for example, was built of 50 equal time steps from ZAMS, with 28 variables computed as a function of solar radius including luminosity, temperature, density, pressure, specific heat at constant volume and sound speed.

Since Bethe's revelation, much research has focused on how the energy produced in the solar reactor is transmitted to space and thence to the target (such as the Earth) under consideration; photons from the core are thought to take perhaps 100,000 years to reach the solar surface because the journey consists of successive random walks. An opportunity to probe the source itself in real time was created by work on the flux of high energy electron neutrinos (ν_e) to Earth, which should amount to some

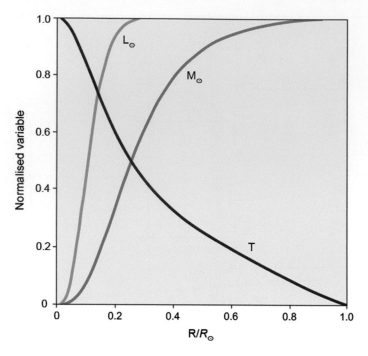

Fig. 5.3 The mass, luminosity and temperature of a standard solar model (SSM) as a function of distance from the Sun's centre (after Bahcall and Pinsonneault[3])

6.5×10^{10}/second/cm^2 at the sunward face of the Earth. Carrying no net electric charge and affected only by the weak force (one of the four fundamental forces of nature) the neutrinos pass through matter almost unimpeded. 'The detection of solar neutrinos is the only experiment that we can think of which could provide *direct* evidence of specific nuclear reactions occurring in the interior of a star'[4].

In the first attempt, in 1968, to compare the predicted neutrino output with reality the arrivals on Earth were ambushed in a 45,000 litre tank of perchloroethylene (C_2Cl_4) in the Homestake Mine in South Dakota by counting per unit time the argon (^{37}Ar) atoms they produced by interacting with the chlorine. The total was 1/3 the number predicted by the current SSM. A deficit persisted when measurement was done in Russia and Italy using gallium instead of perchloroethylene and was reduced, but not eliminated, in Japan using pure water.

In 1997 the sound speeds inferred from helioseismology and those calculated with a standard solar model to predict solar neutrino fluxes agreed closely, indicating that the neutrino predictions were dependable, and they were used to argue somewhat provocatively that solar neutrino experiments require new physics, not revised astrophysics[5]. The neutrino (deficit) problem still mystified science until 2001, when measurements using heavy water (D_2O) at the Sudbury Neutrino Observatory (SNO) in Canada showed that many of the neutrinos produced in the solar interior, which as we saw are of the electron type, oscillate between different neutrino types, namely electron, muon (v_μ) and tau (v_T), by interacting with electrons in the Sun and on their way to the Earth. Muon and tau neutrinos had not been detected at all in the chlorine and gallium systems and only in part in the Japanese experiments with pure water.

Whereas the essential validity of the neutrino modelling was vindicated, the assumption that neutrinos have no mass could thus no longer be justified as they had responded to the presence of matter. But certain issues besides model validity have benefited from the existing measurements. Thus the neutrinos derived from boron 8 (^{8}B) decay in the Sun, though rarer than those due to the p-p cycle, pin down the central solar temperature at the 1% level[20]. The result is relevant to the present chapter also by what it specifies as regards solar luminosity, as the one-solar mass (1 $M_\odot$) model must have the observed radius and luminosity of the Sun at the age inferred for it. What is more, the model's surface composition must match the solar composition ('abundances') that we now observe by spectroscopy[25].

In their role as guides to the solar interior neutrinos do more than validate the model. Once the core reaches a temperature of 6×10^9 K the neutrinos carry off perhaps as much as 6–8% of the Sun's energy while they travel to the surface at the speed of light in 2 s unlike photons, which interact with matter and thus take millennia to walk randomly out to space. More controversially, variations in neutrino flux have been ascribed to the solar (11-year) cycle and to differential rotation of the interior which, if confirmed, would help us understand two major solar features of the modern Sun: sunspots and fluctuations in its luminosity. But employing measurements of neutrino flux to constrain models of the solar interior is difficult because many such models are built on numerous co-varying parameters[1], and as the effects are generally identified statistically, they are open to procedural challenge. In any case there is a gross discordance between the immediacy of the neutrino flux and the years, centuries or millennia over which the photons responsible for solar activity (such as sunspot cycles) operate.

At all events three tests by a team of 124 physicists for high-frequency variations in the ^{8}B solar neutrino flux, a search motivated by the possibility that either the production or the propagation of these neutrinos could be influenced by internal gravity oscillations, yielded no statistically significant signal[1]. The 11-year link is not substantiated in the Homestake data, the longest set[10], or at Super-Kamiokande. But this does not rule out the effect of differential rotation rates deep in the convection zone or at greater depths[48, 51] and thus on further application of neutrino data to probing the present-day solar interior.

Coronal Heating

Side by side with a concern for how the solar interior functions is the need to explain how fusion is translated into solar luminosity in order to understand its variability as a whole and in its constituent parts. When the Sun was thought to glow in response to gravitational contraction, meteoritic impact or some other pre-nuclear explanation, there was little mystery: as Lord Kelvin[34] thought, it (or he, as Kelvin referred to the Sun) is an incandescent liquid simply losing heat—in modern terminology radiant energy, mainly infrared. With acceptance of the Bethe model came the need to translate the energy liberated by fusion. Since the proton-proton chain occurs an estimated 9×10^{37} times per second, and 4.3×10^6 million metric tonnes of matter are converted to energy per second, about 3.9×10^{26} J of energy are released per second and the Sun can run for a further 5 Gyr.

Since it was identified in 1939 the rise in temperature from about 5800 K at the photosphere to over 1–2 million K at the outer margin of the corona has still not been fully explained. Most current proposals hinge on magnetic reconnection or on some kind of wave model[19, 44].

Magnetic reconnection depends on the disruption of magnetic field lines and their attachment to different magnetic poles thereby releasing stored magnetic energy. Reconnection agents at one extreme in scale are solar flares, which may release in excess of 10^{33} erg of energy; at the other are microflares, with energy release of $\sim 10^{27}$ erg and nanoflares ($\sim 10^{24}$ erg), which remain unobserved but could in fact represent the heating mechanism proposed by Parker in 1988[16, 29].

The net impact of the reconnection process has yet to be fully assessed. That it is discontinuous in space as well as time is suggested by the data of the Reuven Ramaty High Energy Spectroscopic Imager (RHESSI), which has found that all the >25,000 X-ray micro-flares it detected in 2002–2008 are confined to solar active regions[18]. And detailed modelling suggests that reconnection may generate heat at best sufficient to maintain the corona at the observed temperature[40]. More may be learnt from NASA's Magnetospheric Multiscale (MMS)[14], a four-spacecraft mission launched in 2015 which investigates magnetic reconnection at the edges of the Earth's magnetosphere.

For its part, wave heating is ascribed either to magnetoacoustic waves or to Alfvén waves which convey energy generated in photospheric granulation until it is dissipated as heat. Of course both could be generated by magnetic reconnection. But, to judge from Transition Region and Coronal Explorer (TRACE) satellite data, acoustic waves have an energy flux of about $3.5 \ 10^2$ erg cm^2/s, far too low for coronal heating[2], and they provide heating in the chromosphere which falls short of the required levels by 90%[33].

A third option, which also explains the threefold structure of the solar atmosphere, and which runs more continuously and unobtrusively than reconnection or waves, tallies with the previous two options to the extent that it assumes 'heating of the solar atmosphere comes from the convection zone'[44]. Thereafter the paths diverge.

In our scheme electromagnetic energy derived from the Sun's convection zone gives rise to ohmic (or Joule) heating (that is to say heating resulting from the passage of an electric current through a conductor) of the chromosphere raising its temperature to $\sim$20,000 K. This triggers Joule-Thomson[32, 49] heating to 250,000 K in the Tansition Region, whereupon plasma expansion into then near-vacuum of space takes over and brings the temperature[53, 54] to 1–2 million K (Fig. 5.4).

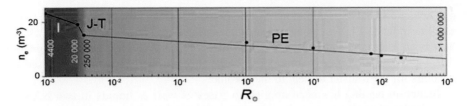

Fig. 5.4 The tripartite scheme for coronal heating: I (induction), J-T (Joule-Thomson effect), PE (plasma expansion), n_e (m^{-3}), electron density, R$_\odot$, radial distance from photosphere. Data points itemised in Vita-Finzi[54]

Electromagnetic energy thus derived from the solar convection zone drives ohmic eddy currents generated by rotating magnetic fields at the summit of Taylor columns (Fig. 5.5). The chromosphere plays the role of the disks that are used to make non-ferrous cookware amenable to heating by induction[57].

The spinning Taylor columns of Fig. 5.5 generate magnetism at the photosphere by a process akin to that associated with a loop in a wire carrying a current. More-over small-scale vortices can give rise to magnetic fields collectively, as shown by the analysis of planar microcoil clusters[15] and consistent with the proposed use of assemblages of spiral coils in order to obtain a uniform magnetic field for wireless power transfer[47]. The photosphere displays vortex flow affecting groups of granules, of which the first attested example was a unit with a diameter of ~5000 km and a lifespan[12] of at least 89 min.

Of course large-scale vortices can arise spontaneously in rotating convection with-out any evident intervention by Taylor columns or granules, and the vortices may give rise to magnetic fields of a size comparable with that of the body in question[26, 27]. Hence our suggestion that the giant convection cells with diameters of ~2 × 10^8 m that have been inferred from the motion of the supergranules[30] could integrate the

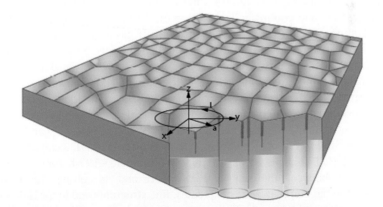

Fig. 5.5 Proposed role of Taylor columns in carrying heat and mass flux to the photospheric surface from the solar interior and generating induction in the chromosphere by their spin

magnetic potential of granulation. At a more modest scale, observations on the SDO satellite indicate EUV cyclones rooted in rotating network magnetic fields, which have been found to be ubiquitous[58] and which contribute 78% of the total magnetic flux. Indeed, solar vortices ('tornadoes') in the chromosphere have been proposed as conduits for the transfer of heat to the corona[55].

Induction heating has been applied to gases as well as liquids in two NASA-supported studies. The first employed a DC arc jet and preheated argon at low pressure but high velocity flowing through a heating duct[9]. In the second an induction plasma torch was successfully operated using hydrogen; the unit was ignited on pure argon but run on pure hydrogen[50] at one atmosphere at 60–160 kW. More recently, magnetic induction has been invoked[17] to account for the acceleration of the solar wind in coronal holes, while coronal heating by ohmic dissipation of electric currents has been shown to be very effective[35]. Structures created in the photosphere by induction are well suited to the release of magnetic energy[21].

Experimental work in this field is dominated by fusion research and in particular variations on the tokamaks model using the deuterium-tritium (hydrogen-2 and hydrogen-3) reaction. In order to attain the requisite startup temperatures ($>10^8$ K) in the plasma, ohmic heating driven by an induced current has to be supplemented, commonly by radio-frequency (RF) heating. However, a major energy requirement in a tokamak is for confining the plasma magnetically, and on the Sun this may be achieved gravitationally.

The distinctive colour of the chromosphere, dominated by hydrogen alpha (Hα) lines (656.3 nm), indicates hydrogen ionisation: optically thin hydrogen plasma becomes 98% ionised at a temperature of 20,000 K, broadly consistent with the range estimated for the chromosphere from Skylab. The temperature decreases from about 8000 K at the base of the chromosphere to 3800 K before rising to ~250,000 K at the Transition Region $\sim 2 \times 10^6$ m above the photosphere.

Hydrogen and helium are two of the three gases (the third is neon) with such a low inversion temperature that above ambient temperature and isolated from their surroundings (isoenthalpic conditions) the response to the Joule-Thomson effect is heating rather than cooling[22, 31, 42] (Fig. 5.6). The effect of pressure change on the temperature of real (as opposed to theoretical) gases was first investigated in the laboratory in France by Gay-Lussac[23] and in the UK by Joule[32, 49]; in their experiments the gas was forced through a porous plug into an evacuated container. A subsidiary issue is the extent to which throttling—or, as Thomson and Joule[49] put it, 'rushing through small apertures'—is critical to the effect: The Joule-Thomson (J-T) effect is a term used here to indicate warming of a real gas subject to free expansion under constant enthalpy if it rises above its inversion temperature (T_i) and cooling if it drops below its T_i. Hydrogen and helium have unusually low T_i s and the inversion point at which μJT changes sign is ~51 K for helium and ~193 K for hydrogen.

Sandwiched between the Sun's chromosphere and its corona, the Transition Region is a thin but complex zone of spicules, fine structures and loops rather than a tidy layer of uniform thickness[45]. The temperature rise begins at 2×10^4 K and attains 2.5×10^5 K over a mere 1×10^5 m. Gold[24] argued that completely ionized gases can attain thermonuclear temperatures by the J-T route. As the ionisation fraction

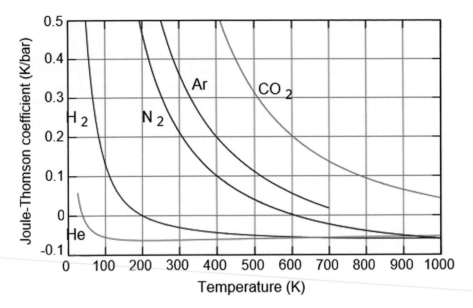

Fig. 5.6 Range of Joule-Thomson coefficients to illustrate low inversion temperature for helium and hydrogen

of pure hydrogen rises steeply at about 10,000 K it now becomes much more susceptible to the J-T effect.

The corresponding change in height is therefore so slight that pressure across the TR is effectively unchanged, and density was consequently thought by Mariska[41] to undergo a 40-fold reduction, but it is much more substantial than 1/40 and consistent with the progressive change in electron density (Fig. 5.4) represented by typical values of 10^{23} m^{-3} for the photosphere, 10^{15} m^{-3} for the TR as a whole, 10^{14} m^{-3} for the base of the corona in quiet regions, 10^{12} m^{-3} at 1 $R_\odot$, 10^{7} m^{-3} at 1 AU, and 10^{6} m^{-3} in the interstellar medium[2, 38, 45]. Empirical electron density profiles derived from white light brightness measurements[39] range from 10^{15} m^{-3} at 0.01 $R_\odot$ to 10^{10} m^{-3} at 10 $R_\odot$ and indicate a similar gradient. Note that novel techniques already permit more subtle differentiation. Hard X-rays, for example, allow neutral gas density in the chromosphere to be evaluated with a resolution of ~150 km; the result is close to a linear decline in number density of over two orders of magnitude over a height of ~10^{6} m above the photosphere[36], by no means inconsistent with our electron plot.

Theoretical studies suggest that the expansion of a plasma into a vacuum or a more tenuous plasma leads to the acceleration of ions greatly above their thermal energy. The effect, which shows an obvious kinship to the J-T effect, was first outlined by Gurevich et al.[28], who claimed that the acceleration could amount to several orders of magnitude and who referred to pertinent experimental studies which had also revealed ion acceleration above thermal levels. The bearing of this effect on space phenomena was made explicit by the interaction of an obstacle with a plasma such as the Gemini/Agena spacecraft and the Moon's wake on the Wind spacecraft.

As things stand there is thus a circumstantial case for continuous though not uniform heating from the photosphere to the upper corona and beyond. A preliminary test of this progression was made by plotting sunspot number (SSN) and average irradiance at 1 min averages as 10-min moving averages[56] over a 6 m period (Jan-June 2012)[52] chosen for its anodyne character within the rising limb of Solar Cycle 24 (Fig. 5.7). The chosen lines represent temperatures spanning 10^3–10^6 K, that is to say from the photosphere as manifested in sunspot number (SSN) through 30.4 nm, corresponding to the upper photosphere/transition zone, 19.3 nm at the upper corona, to 9.4 nm, which represents the corona during a solar flare[43].

Broad agreement was found between the fluctuations displayed by sunspot data for the visible disk (SSN) and irradiance in watts m^{-2} for the coronal lines for He II to Fe XVIII. The broad synchroneity in the timing of peaks and troughs is consistent with heat transmission. The fluctuations ranged from 10^{-4} W m^{-2} for the 30.4 nm line to 10^{-6} W m^{-2} for the 9.4 nm line, that is to say with reduced amplitude but consistent periodicity: prima facie evidence of the radial transmission of a controlling signal. The data span the three orders of magnitude of the problematic temperature ascent.

Why the signal should fluctuate at source is unclear but the parallelism between the coronal and sunspot oscillations points to irregularities in the H/He flux as the primary modulating factor, a useful corrective to the assumption that the core operates with serene efficiency (see Fig. 4.8). A search for radial variations was pursued by referring to 1 min EVE averages for Fe XII and Fe XIV on 20 January 2012 when there appeared to be a discrepancy in the timing of a clear peak amounting to as much as 2.5 h or 0.1 d (Fig. 5.7) Being cooler than the Fe XX component, the Fe XVI emission peaks on average 6 min after the Fe XX and the GOES X-ray peak, and the time delay indicates the cooling rate of the post-flare coronal loops in the volume involved in both the impulsive and the gradual phases. For a long duration event (LDE: C3.2 flare) on 1 August 2010 the delay for the Fe XVI gradual phase peak was reportedly 101 min. In the situation depicted in Fig. 5.8 the hotter line peaked before the cooler by two h. Care had been taken to rule out flare activity by selecting a period following the total decay of a substantial flare (active region 11402) which gave rise to M2.6 and M3.2 class solar flares and a full halo CME: activity levels had apparently fallen to normal background levels as indicated by X-ray flux at 0.5–8.0 nm (Fig. 5.9).

To sum up, the general coherence between solar wind variations and sunspot activity coupled with the modest variability displayed by the various atmospheric zones are consistent with our proposed stepwise heating scheme. The solar wind emerges as the preferred indicator of solar activity. It remains to be seen whether the cosmogenic isotope record, which already spans more than 800,000 years[46], can provide the requisite link to the history of the solar wind on the grounds that it represents its modulating effect on the flux of galactic cosmic rays[6, 51].

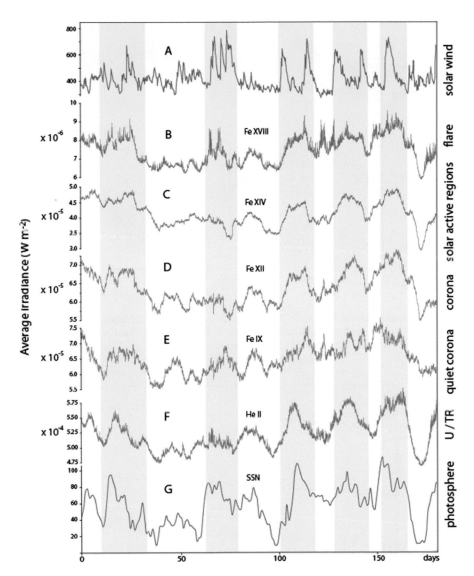

Fig. 5.7 Oscillations during 1 January-1 July 2012 in: A, solar wind, B-F, 5-min averages in five coronal lines, and G, net sunspot number (sources in Vita-Finzi[54])

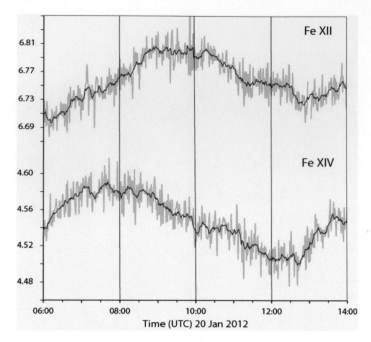

Fig. 5.8 Output of irradiance from Fe XII and Fe XIV coronal lines in 20 January 2012 to illustrate lag in the timing of peak flux with radial position in the corona consistent with heat dissipation independent of flare activity (EVE/SDO data courtesy of NASA)

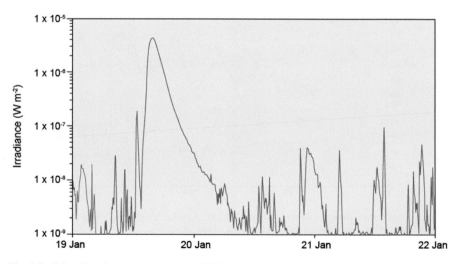

Fig. 5.9 Solar flare from active region 11402 detected by X-ray sensor on GOES-15 spacecraft (courtesy of NASA)

Our focus on hydrogen and helium should not obscure the potential of the proposed mechanism for other solar system gases. The proposed scheme could help to explain heating in other bodies (such as Saturn's moon Titan) which display a radial increase in temperature and decrease in plasma density as well as sustained gas outflow. It may also bear on the thermal evolution of other coronal stars. In any case the coronal temperature and pressure distribution illustrates how astronomical observation can supplant laboratory and computer modelling for initial evaluation of thermochemical processes at extreme settings.

References

1. Aharmin B et al (2010) Searches for high-frequency variations in the ^{8}B solar neutrino flux at the Sudbury Neutrino Observatory. Astrophys Jour 710:540–548
2. Aschwanden M (2009) Physics of the solar corona. Praxis Chichester
3. Bahcall JN, Pinsonneault MH (2004) What do we (not) know theoretically about solar neutrino fluxes? Phys Rev Lett 92:121301
4. Bahcall JN et al (1963) Solar neutrino flux. Astrophys Jour 137:344–345
5. Bahcall JN et al (1997) Are standard solar models reliable? Phys Rev Lett 78:171–174
6. Bard E et al. (1997) Solar modulation of cosmogenic nuclide production over the last millennium: comparison between ^{14}C and ^{10}Be records, Earth Planet Sci Lett 150: 453–462
7. Bethe HA (1939) Energy production in stars. Phys Rev 55:434–456
8. Blackman EG (2018) www.pas.rochester.edu accessed November 2018
9. Boddie W L (1967) An experimental study of radio-frequency induction heating of a partly ionized monatomic gas in a supersonic flow regime. PhD thesis, Rice Univ, Houston TX
10. Boger J, Hahn RL, Cumming JB (2000) Do statistically significant correlations exist between the Homestake solar neutrino data and sunspots? Astrophys Jour 537:1080–1085, https://doi.org/10.1086/309069
11. Borexino Collaboration (2014) Neutrinos from the primary proton-proton fusion process in the Sun. Nature 512:383–386
12. Brandt PN et al (1988) Vortex flow in the solar photosphere. Nature 335:238–24
13. Brown GE, Lee S (2009) Hans Albrecht Bethe. Biog Mem, Nat Acad Sci, Washington DC
14. Burch JL, Moore TE, Torbert RB, Giles BL (2016) Magnetospheric multiscale overview and science objectives. Space Sci Rev 199:5–21
15. Cappelleri D et al. (2014) Towards mobile microrobot swarms for additive micromanufacturing, Int J Adv Robotic Syst 11: 150, https://doi.org/10.5772/58985
16. Cargill P (2013) From flares to nanoflares: magnetic reconnection on the Sun. Astron Geophys 54:3.16–3.20
17. Chertkov AD, Arkhipov YuV (1994) Induction heating of corona and acceleration of solar wind. Proc 1992 STEP/5th COSPAR Coll, Pergamon, Oxford, 117–120
18. Christie S, Hannah IG, Krucker S, McTiernan J, Lin RP (2008) *RHESSI* microflare statistics. I. Flare-finding and frequency distributions. Astrophys Jour 677:1385-
19. de Moortel I, Browning P (2015) Recent advances in coronal heating. Phil Trans Roy Soc A 373:20140269
20. Fiorentini G, Ricci B (2002) What have we learnt about the Sun from the measurement of the ^{8}B neutrino flux? Phys Lett B 526:186–190
21. Gangadhara RT et al (2014) Generation of magnetic structures in the solar photosphere. Astrophys J 788:135
22. Gans PJ (1993) Joule-Thomson expansion. Tcc.iesl.forth.gr/education/local/Labs-PC-II/JT.pdf

23. Gay-Lussac LJ (1807) Premier essai pour déterminer les variations de température qu'éprouvent les gaz en changeant de densité, et considérations sur leur capacité pour le calorique, Mém Phys Chim Soc Arcueil 1:180–203
24. Gold L (1964) Nature of the Joule-Thomson phenomenon in plasmas. Nuovo Cim 34:1371–1380
25. Guenther DB (2010) What is a Standard Solar Model? www.ap.stmarys.ca/~guenther/evolution/what_is_ssm.html
26. Guervilly C, Hughes DW, Jones CA (2015) Generation of magnetic fields by large-scale vortices in rotating convection, arXiv:1503.08599v1 [physics.flu-dyn].
27. Guervilly C., Hughes DW, Jones CA (2017) Large-scale-vortex dynamos in planar rotating convection, arXiv:1607.00824v2 [physics.flu-dyn].
28. Gurevich AV., Pariiskaya LV, Pitaevskii LP (1966) Self-similar motion of rarefied plasma. Sov Phys (Eng Trans) JETP 22: 449–454
29. Hannah I G et al (2011) Microflares and the statistics of X-ray flares. Space Sci Rev 159:263-
30. Hathaway D, Upton, HL, Colegrove O (2013) Giant convection cells found on the Sun. Science 342:1217–1219
31. Hendricks RC, Peller IC, Baron AK (1972) Joule-Thomson inversion curves and related coefficients for several simple fluids. NASA Tech Rep TN-D-6807, Washington DC
32. Joule JP (1845) On the changes of temperature produced by the rarefaction and condensation of air. Phil Mag XXVI:369–383
33. Kalkofen W (2008) Wave heating of the solar chromosphere. Jour Astron' Astrophys 29:163–166
34. Kelvin Lord (Thomson W) (1862) On the age of the Sun's heat. Macmillan's Mag 5:388–393
35. Khomenko E, Collados E (2012) Heating of the magnetized solar chromosphere by partial ionization effects. Astrophys J 747:87
36. Kontar EP, Hannah IG, MacKinnon AL (2008) Chromospheric magnetic field and density structure measurements using hard X-rays in a flaring coronal loop. Astron Astrophys 489: L57-L60
37. Kragh H (2016) The source of solar energy, ca. 1840–1910: from meteoric hypothesis to the radioactive speculation. arXiv:1609.02834v1[physics.hist-ph]
38. Leblanc Y, Dulk GA, Bougeret J-L (1998) Tracing the electron density from the corona to 1 AU, Solar Phys., 183,165–180.
39. Lemaire JF, Stegen K (2016) Improved determination of the temperature maximum in the solar corona. Solar Phys 291:3659–3683
40. Longcope DW, Tarr LA (2015) Relating magnetic to coronal heating. Phil Trans R Soc Lond 373:20140263
41. Mariska JT (1986) The quiet solar transition region. Annu Rev Astron Astrophys 24:23–48
42. Maytal B-Z, Pfotenhauer JM (2013) Miniature Joule-Thomson cryocooling. Springer, New York NY
43. NASA (2013) How SDO sees the Sun. www.nasa.gov/content/goddard (accessed 11 Nov 2016)
44. Parnell CE, de Moortel I (2012) A contemporary view of coronal heating. Phil Trans Roy Soc A 370:3217–3240
45. Priest E (2014) Magnetohydrodynamics of the Sun. Cambridge Univ Press, Cambridge
46. Raisbeck G M et al. (2006) ^{10}Be evidence for the Matuyama-Brunhes geomagnetic reversal in the EPICA Dome C ice core. Nature 444:82–84
47. Shen L et al (2014) Proc IEEE Int Conf Communication Problem-Solving, Beijing, 394–396
48. Sturrock PA, Weber MA (2002) Comparative analysis of Gallex-GNO solar neutrino data and SOHO/MDI helioseismology data; further evidence of rotational modulation of the solar neutrino flux. Astrophys Jour 565:1366–1375
49. Thomson W & Joule JP (1853) On the thermal effects of fluids in motion. Phil Trans Roy Soc London 143: 357–365
50. Thorpe M L, Scammon LW (1969) Induction plasma heating: high power, low frequency operation and pure hydrogen heating. NASA, Washington DC
51. Vita-Finzi C (2013) Solar History. Springer, Dordrecht

52. Vita-Finzi C (2014) Towards a solar system timescale. Astron Geophys 55:4.27–4.29
53. Vita-Finzi C (2016a) The solar chromosphere as induction disk and the inverse Joule-Thomson effect, arXiv:1609.00508v2[astro-ph.SR]
54. Vita-Finzi C (2016b)The contribution of the Joule-Thomson effect to solar coronal heating. arXiv:1612.07943[astro-ph.SR]
55. Wedemeyer-Böhm S et al (2012) Magnetic tornadoes as energy channels into the solar corona. Nature 486:505–508
56. Woods TN et al (2012) Extreme Ultraviolet variability experiment (EVE) on the Solar Dynamics Observatory (SDO): overview of science, instrument design, data products, and model developments. Solar Phys 275:115–143
57. Zhang RY (2012) A generalized approach to plainer induction heating magnetics. MS thesis, MIT
58. Zhang J, Liu Y (2011) Ubiquitous rotating network magnetic fields and extreme-ultraviolet cyclones in the quiet Sun. Astrophys Jour Lett 741, L7

Chapter 6
Solar Magnetism

Abstract The heliosphere—the volume in space that is dominated by the Sun's magnetism and the solar wind—meets the interstellar plasma and magnetic fields at the heliopause, crossed by the space probe Voyager 1 in 2012 some 122 AU from the Sun. Magnetic fields in the Sun itself include those centered on sunspots as well as those that apply to the Sun as a whole and that influence both its internal dynamics and the behaviour of the other bodies in the solar system. The Sun's magnetism is seemingly generated at the tachocline at the base of the convective zone; its reversal roughly every 11 years remains unexplained.

That the Sun exerts some sort of magnetic influence on Earth and thus on intervening space became apparent when a violent solar flare observed by Richard Carrington on 1 September 1859 was followed some hours later by powerful aurorae at exceptionally low latitudes and by major disturbances in telegraph systems and in a magnetometer at the Kew Observatory outside London[6]. A magnetic Sun-Earth link was implicit, but it was not until 2017, after the Cassini, Voyager and Interstellar Boundary Explorer missions had yielded their data, that the shape of the heliosphere could be confidently mapped.

E.N. Parker[40], after whom the 2018 Solar Probe mission to the corona (Fig. 6.1) is named, states that solar dynamo theory is no more than conjecture, that sunspots are the classic example of ignorance still not understood from the basic laws of physics, and that the internal rotation of the Sun established from helioseismology is still without explanation in terms of the hydrodynamics of the convection zone. What follows illustrates these and related scientific mysteries as useful pointers to profounder puzzles.

© Springer Nature Switzerland AG 2018
C. Vita-Finzi, *The Sun Today*, https://doi.org/10.1007/978-3-030-04079-6_6

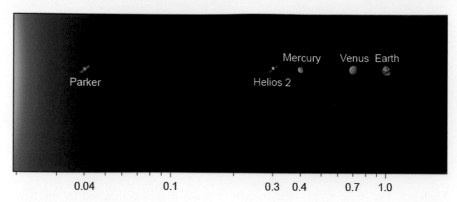

Fig. 6.1 Projected location in AU of the Parker Solar Probe (launched in August 2018) in August 2025. Helios 2 was the spacecraft that has so far made the closest approach, in 1976, when it came to within 43 million km of the Sun

The Heliosphere

The region of space that is dominated by the solar wind and its associated magnetic fields is termed the heliosphere. Its extent and shape are uncertain despite over three decades of satellite exploration. A teardrop shape was long assumed, despite the globular shape implied by the name, on the assumption that movement through space would create a comet-style tail extending for several thousand AUs. In 2015, however, simulations based on data gathered by Voyager 1 suggested instead that the twisted magnetic field of the Sun would drive solar wind jets north and south of the Sun shaped by the interstellar flow to produce a crescent shape with wings extending downwind by no more than 250 AU. That shape (Fig. 6.2) is consistent with images from the energetic neutral atom (ENA) maps produced by the Interstellar Boundary Explorer (IBEX) spacecraft[36].

Yet by 2017 the crescentic image had to be corrected back to something closer to a sphere, albeit one with margins which are constantly in motion as the changeable solar wind beats irregularly against the interstellar medium, which is itself agitated by stellar winds and the fallout from supernovae. Two cameras on the Cassini spacecraft obtained ENA images of the heliosphere and compared the result with the results of a low-energy charged particle experiment on Voyagers 1 and 2 as well as measurements by Voyager 1 of an interstellar magnetic field. The results pointed to a bubble-like heliosphere[13].

The strength of the heliospheric magnetic field in the neighbourhood of the solar system is estimated at only 10^{-6} G but it is still normally sufficient to shield the solar system from GCRs, a notion we have already encountered in connection with solar mass loss (Chap. 4). The existence of a heliosheath and a heliopause (Fig. 6.3), with the implication that the pressure of the interstellar medium is sufficient to confine the heliosphere, was confirmed in dramatic fashion by the instruments on board the Voyager spacecraft.

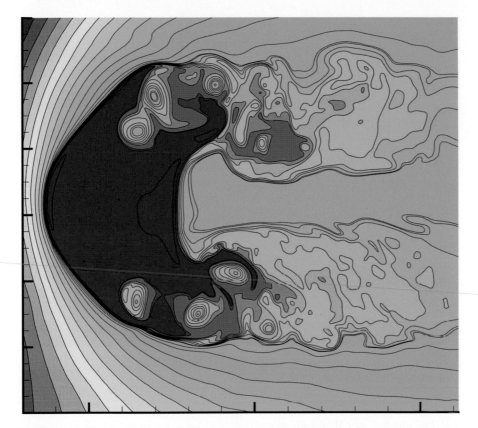

Fig. 6.2 Two-lobe structure of the heliosphere, in contrast with models which postulate a comet-like tail, indicated by simulations which also show that the heliospheric jets are highly turbulent (courtesy of M Opher)

Voyager 2, launched on 20 August 1977, is travelling through the heliosheath. Voyager 1, launched on 5 September 1977, crossed the termination shock on 15 December 2004 at 94 AU and the heliopause on 25 August 2012 at ~121 AU from the Sun. It had detected a sharp increase in high energy, charged—presumably cosmic ray—particles from interstellar space in October 2011–October 2012 and Lyman alpha radiation from the Milky Way galaxy in December 2011. Also in October 2011–October 2012 there was a drastic fall in low-energy (solar wind?) particles[27] (Fig. 6.4).

In 1887 the Michelson-Morley experiment had demolished the notion of a pervasive 'luminiferous aether' by failing to detect any directional effect in the velocity of light. Yet, although the null result helped to clear the way for General Relativity, announced in 1905, Einstein did not rule out some kind of aether: 'space without aether is unthinkable; for in such space there not only would be no propagation of light, but also no possibility of existence for standards of space and time (measuring-

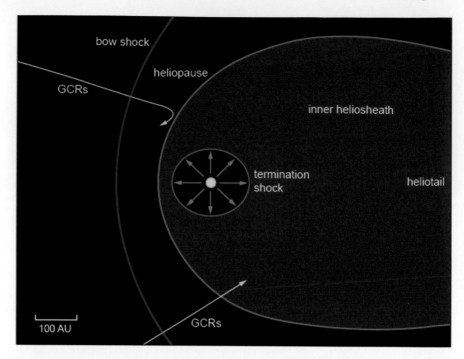

Fig. 6.3 Relationship between bow shock, heliopause and termination shock. Note that some GCRs are deflected by the heliopause

rods and clocks), nor therefore any space-time intervals in the physical sense'[16]. Despite this surprising backsliding, however, the consensus eventually discarded the vapid aether in favour of a severe vacuum populated by a low density plasma of hydrogen and helium as well as magnetic fields, dust, neutrinos, cosmic rays and electromagnetic radiation.

Mapping the magnetic field became one of the subsidiary aims of the Planck space observatory (2019–2013). Planck had as principal target the cosmic microwave background (CMB) at microwave and infra-red frequencies. By detecting in the foreground the polarised emission of interstellar dust grains (the polarisation of starlight by interstellar dust had been recognised in 1949) Planck was able to delineate the structure of the Galaxy's magnetic field, as the grains tend to align at right angles to the field direction (Fig. 6.5). In so doing it revealed the build-up of structures in the Milky Way which might lead to star formation[20]. The effect of the cosmic magnetic fields on distant galaxies, rather than our own, is one of the issues to be explored by the Square Kilometre Array (SKA), due to be completed in Australia and South Africa by 2023. To this end the SKA will use synchrotron emission, Faraday rotation and Zeeman splitting. The Zeeman effect is where a spectral line is split into two or more components in the presence of a magnetic field (Fig. 6.6). It provides a measure of field strength for cold gas clouds[23] just as well as for sunspots.

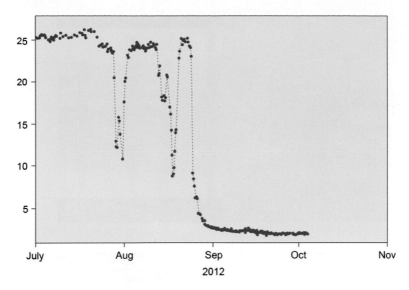

Fig. 6.4 Entry of Voyager 1 into interstellar space in 2012 indicated by abrupt reduction in impact of protons and other particles from the Sun (courtesy of NASA)

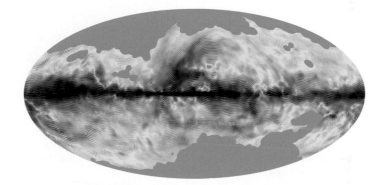

Fig. 6.5 Magnetic map of Milky Way on a Mollwedie projection as seen by the Planck probe. Darker regions indicate regions more strongly polarised; the striations denote the direction of the magnetic field, which is predominantly parallel to the galactic plane (dark band) (courtesy of ESA/Planck and NASA)

Average values for field strength, though very approximate, are instructive. For instance a value of $\sim$50 μG for Interplanetary Space is sometimes cited, with $\sim$5 μG for the Milky Way as a whole but several mG at dense clouds within it, and 1–2 G for the solar poles compared with up to 3000 G at sunspots. The Sun's dipolar magnetic field (discussed in the next section) is twisted by the Sun's rotation into a 'Parker spiral'; it has an amplitude of ~4 G compared with ~0.02–0.05 mG for the radial interplanetary field component at 1 AU[50], that is to say 5 orders of magnitude

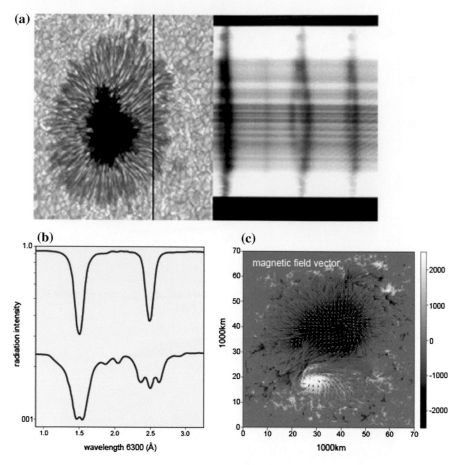

Fig. 6.6 The Zeeman effect. **a** Splitting of spectral line at a sunspot; **b** (upper) iron absorption line split into three components (lower) by strong magnetic field; **c** magnetic field vector at sunspot plotted using Zeeman effect (adapted from Hinode Solar-B website)

greater. The radial component of the heliospheric magnetic field between 80°N and 80°S in 1994 and 1995 measured by the magnetometer on the Ulysses spacecraft[3] was about ~3×10^{-5} G.

The Global Field

Analysis of magnetism on the Sun itself was launched by the discovery of sunspot fields by Hale[28] with strengths of several thousand G. There followed recognition of the 22-year magnetic cycle[29]. Detection of the 'quiet-sun' solar field had to await the development of a sufficiently sensitive device, the photoelectric magnetograph, by

Fig. 6.7 The internal rotation profile of the Sun based on data obtained by the Helioseismic and Magnetic Imager (HMI) aboard the Solar Dynamics Observatory (SDO). The dotted line marks the tachocline at the base of the convective zone within which the rotation rate varies with both depth and latitude (courtesy of NASA)

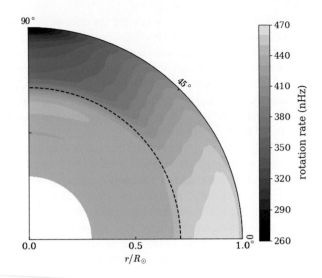

Babcock[44] in 1953, as its strength averages 1 G, only twice as strong as the average field on the surface of the Earth.

Although much has been learnt about magnetic structures in the photosphere, notably the active areas, the Sun's general field is poorly understood. Data gathering by artificial satellites has partially plugged the gap. The Pioneer 5–9 probes between 1959 and 1968 made the first detailed measurements of the solar wind and the solar magnetic field. Many subsequent missions have maintained the tradition. Both the MDI on SOHO (1995–) and the Helioseismic and HMI on the SDO (2010–) targeted (among other things) the magnetism of the solar surface.

The ultimate aim is of course to improve our understanding of the Sun's internal operation, which includes the location and behaviour of the solar dynamo. That some kind of dynamo accounts for the solar field is a longstanding assumption though not always clearly specified. Joseph Larmor[33] linked the motion of sunspots with currents within the Sun and posited a rotational origin for a magnetic field in both the Earth and the Sun, but the explicit application of dynamo theory to the Earth is owed to Walter Elsasser[18], who also considered the relevance in this context of the Sun's convection zone[19].

Helioseismology has sharpened discussion of a dynamo model by quantifying differential rotation as well as the depth of the convective zone[47]. Differential rotation within the photosphere is familiar since early in the history of sunspot studies, with velocities ranging from ~35 days per rotation near the poles to ~25 days per rotation at the equator. That rotation also varies with depth was inferred from a variety of clues, including the imbalance between solar oblateness and surface rotation rates[14]. Helioseismic calculations based on MDI data allow very detailed mapping of such variations and show that, whereas the radiative zone rotates to all effects rigidly, rotation rates within the convective zone vary with depth but more with latitude (Fig. 6.7).

Helioseismology has also opened the door to fresh ways of viewing the magnetic cycle thanks to two crucial findings: poleward flows from the surface to 0.91 $R_\odot$ and 0.82–0.75 $R_\odot$ or deeper, and an equatorward flow between them[52]. In addition there are probably other layers of rotational shear at many levels in the convection zone[48]. The double-cell pattern conflicts with the direction of flow assumed by dynamo models that are used to explain the sunspot cycle, but it can also be used to generate surface magnetic fields which include active regions subject to the internal rotation discussed in an earlier section.

Perhaps most important in fuelling the discussion was the identification of a thin and sharply defined shear layer, soon to be termed the tachocline, at the base of the convection zone, and of a second shear layer near the surface[49]. Thus somewhat vague notions of differential flow as a possible magnetic generator can now be framed in the language of numerical simulation which tries to account for cyclic polarity reversal and sunspot evolution. The tachocline is not a dimensionless contact but a confused layer perhaps 13,000 km (0.019 $R_\odot$) thick. Shear between the convective and radiative zones was long considered an important contributor to the solar dynamo, perhaps in part by analogy with conventional generators (Fig. 6.7).

A central issue remains the mechanism by which the global magnetic field and its ancillaries reverses with some kind of periodicity. In the empirical model for the solar cycle by Babcock[2], because of the Sun's differential rotation the magnetic field oscillates between poloidal (i.e. ~N-S) and toroidal (~E-W) configurations (Fig. 6.8). Parker[38] showed how Coriolis forces in a rotating turbulent medium can recreate the poloidal field from the toroidal one by rendering the turbulence cyclonic[44]. But in one set of numerical simulations the zonal magnetic field bands reversed every 40 years or so[7], strongly encouraging the move away from the classic 11-year explanation but not replacing it by anything quite as digestible.

Magnetic Structures on the Solar Surface

The first and still irreplaceable role of sunspots was to reveal solar rotation and its various contortions. The Schwabe cycle, which still dominates much solar analysis, is their second major contribution, and groups of sunspots together with related features came to be viewed as active areas indicative of subsurface energy sources. The distributed magnetic feature known as the 'magnetic carpet', a network of tens of thousands of N-S poles joined by looping field lines on the quiet sun photosphere (Fig. 6.9), has been ascribed to the recycling of magnetic field by convection on a timescale of 15 min, 1/10 of the flux time for the photosphere[11].

Sunspot formation, however, remains incompletely understood. Sunspots are three-dimensional, dynamic features rather than flat, inert markers, and their interpretation benefits from advances in allied fields such as magnetism, spectroscopy and modelling. The conventional view is that a sunspot is a dark part of the Sun's surface, 1500 K cooler than the surrounding atmosphere, because the presence of a strong magnetic field inhibits the transport of heat via convective motion in the

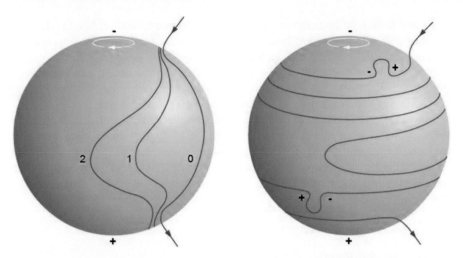

Fig. 6.8 According to the Babcock[2] model, at the approach to solar maximum a toroidal field is generated through differential rotation (left); upwelling in the convective zone drives a toroidal magnetic field through the photosphere and gives rise to sunspots (right)

Fig. 6.9 Magnetic fields in the solar corona. It has been suggested that the magnetic energy in the loops accounts for the elevated temperature of the corona. Image based on data from instruments on SOHO on 19 October 1996 (courtesy of ESA)

Sun[5]. But it may be objected (as was over 30 years ago) that, although the inhibition of convective transport beneath the photosphere may indeed produce a dark area surrounded by a bright ring, the mean temperature beneath the spot will be above normal and the enhanced gas pressure will disperse rather than concentrate the magnetic field[39]. Hence sunspots represent enhanced rather than inhibited energy transport and suppressed convection cannot be their primary cause.

Enhanced energy output is in fact consistent with active region AR 2665, monitored by the SDO in July 2017. In the course of 13 days in direct view the region produced a medium-sized (M1) flare, another medium-sized (M2) flare which was accompanied by a CME, and a SEP. In 2014, AR 2192 (Fig. 6.10 a and b), with its 'behemoth sunspot', and 125,000 km across, discharged 6 X-flares, 29 M-flares and 79 C-flares[45].

Impact on the Earth

A distinction is sometimes drawn between those components of the Sun's magnetic field with field lines which remain attached to the Sun and those with field lines which are dragged out into the heliosphere. The distinction is by no means sharp, notably when coronal mass ejections are unleashed, but it is a useful reminder that the solar wind embodies magnetic field which may slyly penetrate deep into the Earth's and other planetary magnetospheres.

There are various sources of variability in Sun-induced geomagnetic effects, including those that are mediated by the solar wind and the Interstellar Magnetic Field (IMF), annual variability resulting from the Earth's orbit, which bring it to different solar latitudes, and solar rotation, which yields 27-day periodicities[40].

CMEs, as noted in Chap. 4, may amount to 10^{11}–10^{13} kg of plasma and can attain velocities of over 2000 km/s; they are generally more frequent near solar maximum and average ~20 p.a.[32]. The delivery of CMEs is an aspect of space weather which is viewed with alarm by virtue of the damage it could inflict on power installations and communications. For example a CME fired on 9 March 1989 created a geomagnetic storm which damaged the power grid of the Province of Quebec, affected the performance of satellites in polar orbit, and created aurorae as far south as Florida.

Other planets, whether in our solar system or orbiting other stars, are open to this form of solar aggression. On 23 December 2006 the Venus Express probe observed the impact of a CME on Venus, which like Mars lacks a detectable internal magnetic field and is protected solely by an ionosphere in which advecting interplanetary field lines may induce a magnetosphere[51]. The CME was slow (mean 371 km/s when solar wind mean was 361 km/s). Even so it contributed, as did previous CMEs and solar wind pressure pulses, to atmospheric loss at Venus and Mars (see Fig. 6.12) as shown by ionospheric outflow[12, 15].

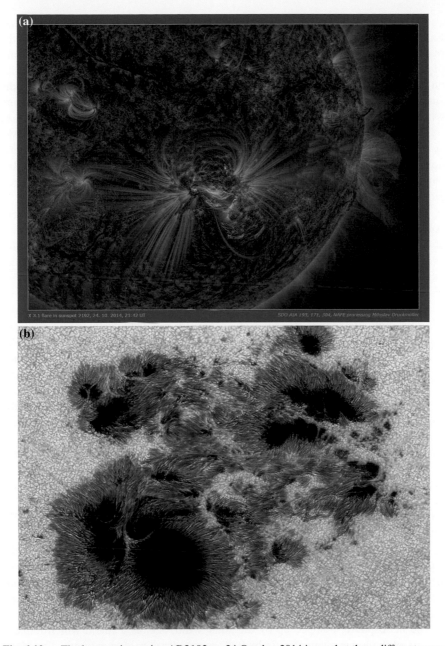

X 3.1 flare in sunspot 2192, 24. 10. 2014, 21:42 UT   SDO AIA 193, 171, 304, NAFE processing Miloslav Druckmüller

Fig. 6.10 **a** The large active region AR2192 on 24 October 2014 imaged at three different wavelengths of extreme ultraviolet light (blue, white and red). Emission from highly ionized iron and helium atoms traces magnetic field lines through the outer chromosphere and corona; the cooler photosphere appears dark at extreme ultraviolet wavelengths (courtesy of NASA). **b** The large sunspot group AR2192 (copyright Randall Shivak and Alan Friedman, with permission)

The magnetised plasma of the solar wind conveys magnetic field, and the near-Earth IMF has an annual average[37] for 1990–2010 of $4 - 9 \times 10^{-5}$ G. A measure of the resulting net magnetic flux is the aa index, which is a 3-hourly magnetometer measure of geomagnetic activity based on readings from roughly antipodal observatories on the Earth, currently Hartland in the UK and Canberra in Australia[9, 36]. Periodic variations in the aa index may result from some kind of orbital cyclicity[42, 44] including where the Earth intersects high-speed solar wind streams[25] that can attain ~750 km/s.

But, at least as regards the 22-year cycle, there are grounds for accepting the IMF as simply part of Sun's internal metabolism[8, 41]. The doubling in the solar magnetic field between 1901 and 1999[35] is more provocative as it coincided with rising global temperatures during 1880–1990[9], reminiscent of a current such linkage between global temperature and GCR flux[6].

The CME associated with the Carrington event of 1859, and advertised by stupendous aurorae, induced currents which disrupted the telegraph network over thousands of kilometres and locally supplanted battery power to maintain communication for two hours[26]. Any repetition of the geomagnetic storm will doubtless prove as decorative and more destructive[24].

The aurorae—borealis and australis—surround the Earth's magnetic poles in an oval 65–70° north or south (Fig. 6.11) and expand if the solar wind is especially vigorous, notably when in the guise of a CME. Aurorae are produced by electrons from the solar wind, which are accelerated along magnetic field lines when the solar wind transfers energy into the magnetosphere and thence into the ionosphere; atoms of various constituents of the upper atmosphere are excited and then emit the energy as photons. Auroral 'curtains' identify magnetospheric field lines.

The colours on display depend among other things on height, and therefore the relative concentration of one or other gas. For example, green, the most prominent auroral colour, is derived from oxygen (557.7 nm) which is relatively abundant at 100 km, a visual effect enhanced by the sensitivity of human vision peaking at about 560 nm, doubtless of evolutionary benefit because green dominates photosynthesis. Red occurs mainly above 300 km and is produced by excited atomic oxygen (630 nm). Molecular nitrogen and ionized molecular nitrogen at lower altitudes give rise to various colours dominated by blue (428 nm). Aurorae may also be present invisibly at ultraviolet, x-ray, infrared and radio wavelengths.

The electric field-aligned currents that accompany CMEs are named after the explorer Kristian Birkeland. They carry 10^5–10^6 amps and raise the temperature of the upper atmosphere by Joule heating by hundreds of degrees. The resulting atmospheric expansion by 1000% or more is likely to alter satellite orbits inconveniently and unpredictably[1, 4].

The heliosphere, pumped up by the solar wind, acts as a shield which reduces the flux of the lower energy GCRs by 90% especially at solar maximum. The effect was initially recognised when a precision cosmic-ray meter in Maryland, in the absence of barometric or temperature fluctuations, displayed a '24-h wave in apparent cosmic-ray intensity, with an amplitude of 0.17% of the total intensity'[21]. This modest observation blossomed into the revelation that the flux of galactic (as opposed to solar) cosmic rays, consisting mainly of protons and alpha particles emanating from

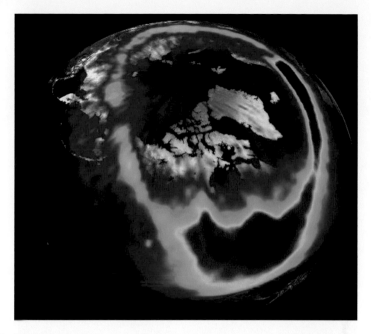

Fig. 6.11 Polar/VIS satellite image of the aurora over the USA on 16 July 2000

outside the solar system and endowed with energies over the vast range 1 GeV–10^8 TeV, was in anti-correlation with the solar ~11-year period[22]. The effect has now been shown to vary with the solar wind and with the IMF at different timescales ranging from days to millennia[46], through CMEs at solar maximum and high-speed streams at solar minimum[10].

Not so benevolent is the impact of the solar wind on Mars (Fig. 6.12) as it is engaged in stripping ions from the planet's upper atmosphere. NASA's Mars Atmosphere and Volatile Evolution (MAVEN) mission has found evidence that, thanks to sputtering driven by the solar wind combined with radiation, 65% of the argon that was ever in the atmosphere has been lost to space. CO_2 and other gases were also lost this way; gas is still being lost at a rate of about 100 g/s, seemingly trivial but cumulatively serious[31].

Similarly Mercury, devoid of a protective magnetic field, is subjected to ion sputtering, and displays an exosphere composed of atoms dislodged from its surface by the solar wind as well as by micrometeoroids. It has even been suggested that the solar wind is responsible for Mercury's puny magnetic field because it deflects charged particles from the magnetosphere, which is magnetically coupled to the internal dynamo, and thus weakens it[30].

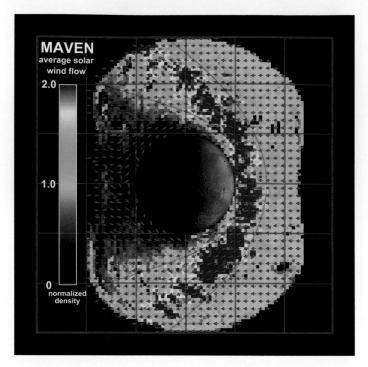

Fig. 6.12 Average solar wind flow at Mars as observed by the Mars atmosphere and volatile evolution mission (MAVEN) spacecraft. The red colour corresponds with higher observed solar wind densities (courtesy NASA/GSFC and the MAVEN team)

References

1. Anderson BJ et al (1998) UARS observations of Birkeland currents and Joule heating rates for the November 4, 1993, storm. J Geophys Res 103:26323–26335
2. Babcock HW (1961) The topology of the Sun's magnetic field and the 22-year cycle. Astrophys J 133:572–587
3. Balogh A, Marsden RG, Smith EJ (2001) The heliosphere near solar minimum. Praxis, Chichester
4. Birkeland K (1908) The Norwegian Aurora Polaris Expedition 1902–1903. Aschehoug, New York
5. Bray RJ, Loughhead RE (1979, first pub 1964) Sunspots. Dover, New York
6. Carrington RC (1859) Description of a singular appearance seen in the Sun on September 1, 1859. Mon Not R Astr Soc 20: 13–15
7. Charbonneau P and Smolarkiewicz PK (2013) Modeling the solar dynamo. Science 340:42–43
8. Cliver EW, Boriakoff V, Bounar KH (1996) The 22-year cycle of geomagnetic and solar wind activity. J Geophys Res 101:27091–27110
9. Cliver EW, Boriakoff V, Feynman J (1998) Solar variability and climate change: geomagnetic aa index and global surface temperature. Geophys Res Lett 25:1035–1038
10. Cliver EW, Richardson IG, Ling AG (2013) Solar drivers of 11-year and long-term cosmic ray modulation. Space Sci Rev 176:3–19
11. Close RM et al (2004) Recycling of the solar corona's magnetic field. Astrophys J 612:L81–L84a

12. Collinson GA et al (2015) The impact of a slow interplanetary coronal mass ejection on Venus. J Geophys Res Space Phys 1230:3489–3502
13. Dialynas K et al (2017) The bubble-like shape of the heliosphere observed by Voyager and Cassini. Nature Astr 1:0115
14. Dicke RH (1964) The sun's rotation and relativity. Nature 202:432–435
15. Edberg NJT et al (2011) Atmospheric erosion on Venus during stormy space weather. J Geophys Res 116:A09308
16. Einstein A (1922) Sidelights on relativity. Methuen, London
17. Elliott JR and Gough DO (1999) Calibration of the thickness of the solar tachocline. Astrophys J 516:475–481
18. Elsasser WM (1947) Induction effects in terrestrial magnetism. Part III. Electric modes. Phys Rev 72:821–833.
19. Elsasser WM (1956) Hydromagnetism. II. A review. Am J Phys 24:85–110
20. ESA (2016) The magnetic field along the galactic plane.
21. Forbush SE (1937) On diurnal variation in cosmic-ray intensity. J Geophys Res 42:1–16
22. Forbush SE (1954) World-wide cosmic ray variations, 1937–1952. J Geophys Res 59:525–542
23. Gaensler BM, Beck R, Feretti L (2004) The origin and evolution of cosmic magnetism. arXiv: astro-ph/0409100v2
24. Giegengack R (2015) The Carrington coronal mass ejection of 1859. Proc Am Phil Soc 159:7–19
25. Gosling JT et al (1976) Solar wind speed variations: 1962–1974. J Geophys Res 81:5061
26. Green J et al (2006) Eyewitness reports of the great auroral storm of 1859. Adv Space Res 38:145–154.
27. Gurnett DA et al (2013) In situ observations of interstellar plasma with Voyager 1. Science 341:1489–1492
28. Hale GE (1908) On the probable existence of a magnetic field in sun-spots. Astrophys J 27: 315–343
29. Hale GE et al (1919) The magnetic polarity of sun-spots. Astrophys J 49:153–178
30. Heyner D et al (2011) Evidence from numerical experiments for a feedback dynamo generating Mercury's magnetic field. Science https://doi.org/10.1126/science. 1207290
31. Jakosky BM et al (2015) Initial results from the MAVEN mission to Mars. Geophys Res Lett 42:8791–8802
32. Jian L et al (2008) Evolution of solar wind structures from 0.72 to 1 AU. Adv Space Res 41:259–266
33. Larmor J (1919) Possible rotational origin of magnetic fields of sun and earth. Elec Rev 85:415
34. Lockwood M, Stamper R, Wild MN (1999) A doubling of the Sun's coronal magnetic field during the past 100 years. Nature 399:437–439
35. Mayaud PN (1972) The aa indices: a 100-year series characterizing the magnetic activity. J Geophys Res 77:6870–6874
36. Opher M et al (2015) Magnetized jets driven by the sun: the structure of the heliosphere revisited. Astrophys J Lett 800:7
37. Owens MJ, Forsyth RJ (2013) The heliospheric magnetic field. Liv Rev Solar Phys 10:5
38. Parker EN (1955) Hydromagnetic dynamo models. Astrophys J 122: 293–314
39. Parker EN (1974) The nature of the sunspot phenomenon. Sol Phys 37:127–144
40. Parker EN (2009) Solar magnetism: the state of our knowledge and ignorance. Space Sci Rev 144:15–24
41. Rasinkangas R (2008) Geomagnetic activity. Wiki.oulu.fi, retr 24 June 2018
42. Rosenberg RL, Coleman PJ Jr (1969) Heliographic latitude dependence of the dominant polarity of the interplanetary magnetic field. J Geophys Res 74:5611–5622
43. Russell CT, McPherron RL (1973) Semiannual variation of geomagnetic activity. J Geophys Res 78:92–108
44. Stenflo JO (2015) History of solar magnetic fields since George Ellery Hale. arXiv:1508. 03312v1 [astro-ph.SR]

45. Thalmann JK et al (2016) Exceptions to the rule: the X-flares of AR 2192 lacking coronal mass ejections. ASP Conf Ser 504:203–204
46. Thomas SR, Owens MJ, Lockwood M (2014) Galactic cosmic rays in the heliosphere. Astron Geophys 55:5.23–5.25
47. Thompson MJ (2004) Helioseismology and the Sun's interior. Astron Geophys 45:4.21–4.25
48. Thompson MJ et al (2003) The internal rotation of the Sun. Ann Rev Astron Astrophys 41:599–643
49. Toomre J, Thompson MJ (2015) Prospects and challenges for helioseismology. Space Sci Rev 196:1
50. Wang Y-M, Sheeley NR Jr (2003) Modeling the Sun's large-scale magnetic field during the Maunder Minimum. Astrophys J 591:1248–1256
51. Zhang TL et al (2008) Induced magnetosphere and its outer boundary at Venus. J Geophys Res 113:E00B20
52. Zhao J-W et al (2013) Detection of equatorward meridional flow and evidence of double-cell meridional circulation inside the Sun. arXiv:1307.8422v1[astro-ph.SR]

Chapter 7
Sun and Weather

Abstract Our grasp of the solar factor in weather remains imperfect, to the detriment of forecasting and palaeoclimatic analysis, but there is progress on several novel fronts, and the study of exoplanet atmospheres provides useful pointers. The influence of galactic cosmic rays on cloud cover is being assessed experimentally. Phytochemistry allows the broadbrush lessons of ecology to be refined. Ozone monitoring and palaeohydrology shed light on the impact of solar UV on the global circulation.

Early in his *Lectures on Physics* Richard Feynman, after a confident gallop through the key notions of stellar astronomy, asserted that 'the theory of meteorology has never been satisfactorily worked out by the physicist' and concluded 'quickly we leave the subject of weather'[11]. More positively G. C. Abbot, director of the Smithsonian Institution's Astrophysical Observatory, in the face of scepticism both from meteorologists and from statisticians, had devoted the years between 1895 and 1944 to improving and refining the proposition first advanced by his predecessor S. P. Langley that short-term cyclic variations in the Sun's radiation influence weather in a predictable way[7].

Putting a value on the Sun's influence remains problematic, witness the tenor of successive introductory sections of the Intergovernmental Panel on Climatic Change[20, 21], which was created in 1988 to provide the world with an objective view of climate change and its impacts. The Physical Science Basis of the Fifth Assessment Report, written or reviewed by 831 experts, and referring to 9200 papers, states with regard to solar irradiance (§ 8.4.1) that there was low confidence in the exact value of solar radiative forcing, and that in any case differences between TSI measurements have little impact on climate simulations because they are dwarfed by uncertainties in cloud properties[28]. An earlier account[20] had already noted that confidence level in assessments of the solar contribution to global mean radiative forcing was 'very low'.

It may be that the enquiry is asymmetrically framed, rather as in the hoary nature-nurture debate. The solar factor is reductionist, generally being expressed in terms of wavelengths, frequencies and the like. Although the atmosphere too is analysed daily in terms of local temperature, pressure and so on, and some influential assessments of climate change focus on a single globally averaged surface temperature, weather on Earth—and climate all the more so—are commonly discussed in terms primarily of air masses, fronts, temperature range, wind regimes and other such aggregate or derivative properties; and it was turbulent flow that unnerved Feynman.

Galactic Cosmic Rays

In Chap. 5 it was argued that the solar wind is possibly the best guide to the solar pulse. Nevertheless its direct influence on terrestrial weather is subtle. A possible example is the Mansurov effect, whereby the interplanetary magnetic field—the magnetic field swept towards the Earth by the solar wind—leads to pressure anomalies above the Earth's polar regions which in a matter of days promote changes in cloud cover[24].

Some kind of role for the solar wind in the Earth's climate has been mooted for over half a century. The initial observation[29] was that solar-cycle modulation of cosmic rays had a large stratospheric and tropospheric effect. It was followed some years later[8] by the recognition that cloudiness could be an important product of 'solar-related fluctuations'; that this might be linked to ionisation near the tropopause by galactic cosmic rays, the only geophysical phenomenon not connected with processes operating in the upper atmosphere that was negatively correlated with solar activity; and that the process required sulphate aerosol particles to serve as cloud condensation nuclei. The importance of cloud type and cloud height was also stressed and quantified: a 0.5 km increase in cloud height, for example, would increase surface temperature by the equivalent of 2% in the solar constant.

All these matters have come to the fore in subsequent discussions of the solar factor. The subject gained impetus from a report that the Earth's low cloud cover and GCR flux change in unison in the course of a solar cycle (1984–1994)[40]. Changes in GCR flux over as brief a period as 4 days where suitable precursor conditions were met have likewise been shown to prompt statistically significant changes in mid-latitude cloud cover and a consequent change in surface level air temperature[23]. Now it is accepted by all parties that the solar wind partly shields the Earth from the flux of galactic cosmic rays (Fig. 1.5), an expansion of the planet-wide observation[12] that GCRs declined drastically immediately after a solar storm. What proved less generally palatable was the implication that, as low clouds promote cooling by their albedo effect, a substantial part of present-day global warming could be explained indirectly by solar activity.

The key mechanism in the controversial proposal is the formation by GCRs of new aerosols—a suspension of fine droplets—which would act as cloud condensation nuclei. The notion was seen to undermine, however unintentionally, international efforts to combat the generation of greenhouse gases, notably the CO_2 produced by the burning of fossil fuels. Its proponents were at best denounced as reckless, at worst as disingenuous allies of big oil.

In view of the need to probe a poorly understood process crucial to climatology[17], and doubtless energised by the political implications of any serious challenge to

Fig. 7.1 View of the Cosmics Leaving Outdoor Droplets (CLOUD) experiment facility at CERN (courtesy of CERN)

the IPCC position, experimental verification was evidently needed to complement empirical evidence. The outcome of these pressures would at other times have seemed extravagant as well as surprisingly imaginative. In 2006 CERN set up CLOUD (Cosmics Leaving Outdoor Droplets), a facility first proposed in 1998 for investigating cloud microphysics using a beam from a particle accelerator (a proton synchrotron) to simulate cosmic rays inside a cloud chamber (Fig. 7.1). By 2009 it was fully operational.

Among its findings by 2016 was that the dominant source of aerosol particles in preindustrial times was organic vapours emitted by trees and oxidised in the atmosphere and that they have an important effect on cloud formation. This would provide a valuable link with a well established strand of palaeoclimatic research. Much of the published organic evidence for past climates consists of pollen and other kinds of plant remains, skeletal material, and geochemical indicators of life. None of these is routinely analysed for short-term change, as the main thrust of the work is generally climatic history at a regional scale over decades, centuries or millennia.

Nevertheless components of the mosaic can sometimes deliver useful clues to past solar activity. Tree ring width, for example, often owes more to water availability than to any more explicitly solar factor, but ring width in the birch *Betula ermanii* in central Japan has been found to be positively correlated with the duration of insolation during July at the timberline[43] at 2400 m. Insolation does not of course rule out other controlling variables. Thus isotopic analysis of tree rings by laser ablation may reveal the intra-seasonal response to different environmental variables[38] and make available tree-ring archives hitherto analysed for ^{14}C geochronology or for identifying drought years. But even then the links with irradiance remain elusive especially as tree rings form in only part of the year and are dependent on stored rather than newly assimilated carbon and thus with fossil rather than current isotopic (^{13}C) climatic signals[14].

UV-B (280–315 nm) radiation has been found to play a regulatory role in plant growth and development, but the consequences of plant response at the molecular level remain uncertain. An interesting exception is the stability of UV-absorbing compounds in plant tissues[2] but it may be difficult to distinguish between the solar flux and changes in stratospheric ozone. Similarly, the anti-correlation of solar activity with the flux of galactic cosmic rays as cloud seeding agents has inspired the suggestion that interannual variations in growth rings in Sitka spruce (*Picea sitchensis*) in northern Britain reflect changes in the production of cloud condensation nuclei by GCRs which in their turn favour photosynthesis in the forest canopy[6].

If a photochemical reaction leads to the chemical transformation of an organic molecule, we have the promise of uncovering the signature of solar radiation in living things or their fossil remains. UV-B absorbing compounds are found in modern and fossil pollen spores and seed coats, raising the prospect of reconstructing UV and ozone changes over the centuries, millennia and possibly longer[30]; all the more intriguing given that there are forms of life which respond to light with great rapidity and with great sensitivity. For example the sporangiophore (stalk) of the fungus *Phycomyces blakesleeanus* responds by bending to far-UV, near-UV and blue light. For blue light (450 nm) its threshold is about 10^{-6} J/m^2 in 10 s, 100–1000 times as sensitive as oat or corn seedlings[10]. Whether the linkage can be inverted, as we once usefully exploited the canary's sensitivity to mine damp, is uncertain.

At the very least the *Phycomyces* story has a cautionary element. The sporangiophere can adapt to light levels ranging over a factor of 10 billion, very like the human eye. What is more, the flavin that makes this sensitivity and adaptability possible in *Phycomyces* and humans was found after a century of study to be only one of several photoreceptive pigments including some that, unlike flavin, could account for the organism's sensitivity to red rather than blue light. A fungus with the 'unseemly' life habit of growing on animal droppings[10] thus underlines the obstacles faced by heliobiology, especially the plasticity of organic response to solar fluctuations.

More important for the cosmic ray dispute were two CLOUD observations: that nucleation on acid and ammonia aerosol particles in the lower atmosphere occurs in the chamber at 1/10–1/1000 the observed rate in nature whereas in the mean troposphere and above it is enhanced as much as tenfold by cosmic ray ionisation; and that ionisation of the atmosphere by cosmic rays accounts for nearly 1/3 of all particles formed[22, 41]. Although the aerosols produced in the experiments were too small (~1.7 nm) to affect clouds, the experiments had confirmed the general GCR thesis.

The authors of the 5th IPCC report, however, were not persuaded, and concluded, after conceding that 'long-term variations in cloud properties are difficult to detect

... while short term variations may be difficult to attribute to a particular cause', that 'there is *medium evidence* and *high agreement* that the cosmic ray-ionization mechanism is too weak to influence global concentrations of CCN (cloud condensation nuclei) or droplets or their change over the last century or during a solar cycle in any climatically significant way'. The possible role of the global electric circuit in linking the cosmic ray flux to cloudiness[47] was likewise dismissed as being imperfectly understood and lacking the requisite evidence of climatic significance[3]. Yet theoretical and experimental data have since shown that ionization promotes the growth of aerosols into CCNs large enough (>20 nm) to influence cloud properties[42].

One suggestion is that the requisite flux of GCRs is supplied by the explosion of supernovae at a rate which, though modulated in the long term by the Sun's passage through the spiral arms of the Galaxy at a tempo measured in 100 M year[34], in the short term is effectively uniform: witness the results of a study using ^{26}Al to show that on average a massive star explodes in the Milky Way Galaxy every 50 years or so (see Chap. 3)[9]. ^{60}Fe, produced by Type II deflagration supernovae (SNe), has been proposed as a means of correlating the history of various solar system bodies by identifying debris from a supernova dated 2.1–2.8 M year[51]. This would seem to run counter to the evidence for a ~50 year periodicity in supernova impact on Earth, but it remains to be seen whether the GCRs invoked in the CLOUD studies, unlike the debris or spallation products of SNe explosions, leave an isotopic trace.

UV

So much for the solar wind as GCR gatekeeper. What of the traditional view of a direct solar factor in weather, namely solar irradiance? Authoritative reviews of the subject[15, 37] have emphasised solar variability, doubtless in response to current concerns over climatic change but perhaps also because likely causes and effects are all subject to change. In any case a moving target in some settings may be easier to hit than a stationary one. The position taken here is analogous to a medical diagnosis where case history is allowed to colour but not obscure the current symptoms once these are securely identified within an accepted range of values.

Solar irradiance is a reasonable starting point, with some grudging allowance made for variation between the flux at sunspot maximum and at sunspot minimum and hence requiring a blurred target of a dozen years or so (e.g.[15]), grudging because the Schwabe cycle, though the most self-evident modulation of the Sun's activity, is not necessarily the most significant (see Chap. 4). But there is also a need to focus on narrowly defined portions of the solar spectrum however contrary this may be to the spirit of ecology.

Solar UV wavelengths account for about 30–60% of TSI variations[28], bearing in mind that the X-ray fraction (0.1–100 nm)(Fig. 7.2) can increase a thousandfold during a major flare[13]. More important in the present context, they provide the bulk of

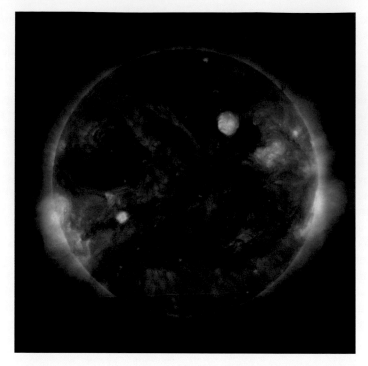

Fig. 7.2 The Sun in X-rays. High-energy X-rays from Nuclear Spectroscopic Telescope Array (NuSTAR) are shown in blue; low-energy X-rays from Hinode spacecraft are green; and extreme UV light from SDO is yellow and red (courtesy of NASA)

the energy and ionisation affecting the upper atmosphere. Figure 7.3 shows average daily insolation measured by several Earth-observing satellites; Fig. 7.4 is a sample of the UV-B (250–320 nm) flux.

UV radiation breaks up oxygen molecules into their constituent atoms which then combine with other oxygen molecules to form ozone (O_3). The ozone in turn modulates the impact of the UV on the atmosphere: its absorption of radiation differs according to altitude and wavelength, being most pronounced in the stratosphere and mesosphere at λ 200–400 nm whereas absorption at longer wavelengths is important below 25 km and at λ < 200 nm above 60 km[31]. Indeed many accounts of the UV flux somewhat prematurely use ozone content as proxy for UV flux.

The rapidity with which ozone levels can change is illustrated by the change in total column ozone, that is to say the O_3 content in a vertical column running through the atmosphere, between two days (Fig. 7.5). The Dobson units widely used for such assessments refer to the differential absorption of solar UV by the ozone layer. The measurement is made using a monochromator, a device which makes it possible to select a narrow band of wavelengths, in the example illustrated the Ozone Measuring Instrument (OMI) on board the Ausra spacecraft.

The density of the thermosphere—the zone in the atmosphere 500–1000 km above the Earth's surface within which UV can provoke ionisation—responds to the

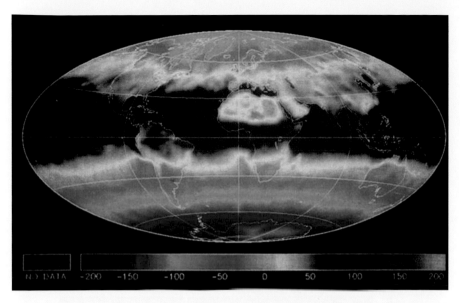

Fig. 7.3 Net radiative flux—absorbed solar flux minus emitted flux—in W/m^2 (ERBS and NOAA9 data courtesy of NOAA)

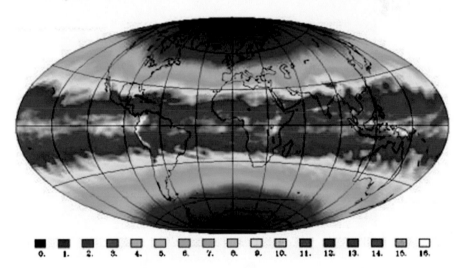

Fig. 7.4 Clear-sky UV B irradiance index for 26 March 2001 (courtesy of Koninklijk Nederlands Meteorologisch Instituut (KNMI) and ESA)

27-day rhythm of the solar EUV (10–120 nm) signal[45]; so does the ozone content of the stratosphere[18] and, apparently without help from GCRs, so did the periodicity in cloud cover over the western Pacific in 1980–2003[44]. The 27-day cycle also appears to control the roughly biennial oscillation (QBO) in equatorial winds[32].

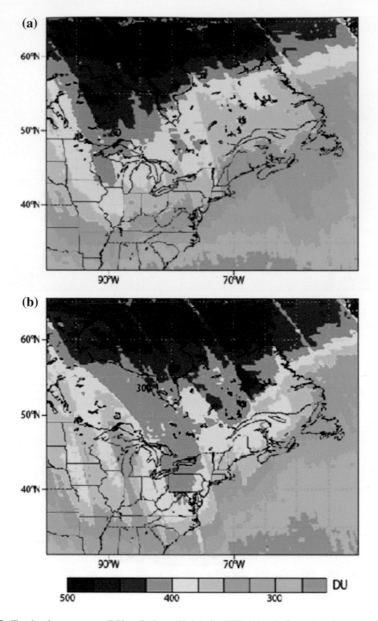

Fig. 7.5 Total column ozone (DU = Dobson Units) for NW Atlantic for **a** 1–2 January 2012 and **b** 2–3 January 2012, measured by EOS Aura Ozone Measuring Instrument (OMI) (data courtesy of NASA)

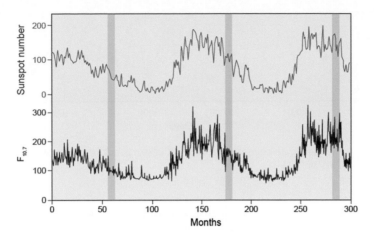

Fig. 7.6 $F_{10.7}$ and solar sunspot number (SSN) for solar cycles 20–22. Vertical bands indicate years with storm track maxima in the North Atlantic for winter (DJF) in 1968–1969 during declining solar cycle phase (data courtesy of NOAA and NASA)

There is a close match between fluctuations in $F_{10.7}$, the Sun's emission at the radio wavelength 10.7 cm which is a faithful proxy of solar activity, and daily sunspot number (SSN) (Fig. 7.6). The match lends support to the view that EUV output is modulated by the passage of sunspots and other active regions across the rotating solar disk so that, even though sunspot *number* explains no more than 42% of total solar irradiance (TSI)[26], the crucial UV wavelengths are involved in the 11-year Schwabe periodicity.

Moreover our specimen plot of average atmospheric temperature shows a strict annual periodicity (Fig. 7.7) which is consistent with heating of stratospheric ozone. In other words the solar signal can be detected with an 11-year period and also with a simple annual tempo. As we have seen, whereas the Sun's convective zone rotates more rapidly at the poles then at the equator, the radiative zone rotates uniformly with regard to latitude[46]. Thus solar irradiance may be modulated by rotation of the radiative zone or even of the solar core rather than by the transit of activity centres in the photosphere[39], a notion which is supported by the presence of the same period in ACRIM measurements and by indications of rotational rigidity in the corona, a major source of solar UV[50]. The next question must surely be whether these associations apply to fluctuations measured in tens hundreds or thousands of years, such as the 'grand solar minima' including the Maunder Minimum of 1645–1720, that are conventionally associated with sunspot history.

A wholly static viewpoint is of course impossible with something as flighty as the atmosphere, but the familiar weather map shows the potential for analysis, prospective and retrospective, provided by synoptic presentation in the light of experience. But in order to test the outcome reliance is generally placed on some kind of model of the atmospheric circulation. One such model which has undergone more than 25 years of refinement is the Goddard Institute for Space Studies General Circulation Model

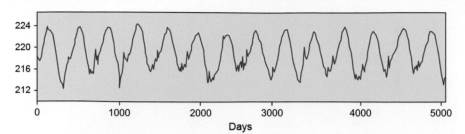

Fig. 7.7   Zonal average stratospheric temperature in K at 30 mb for 23 February 1979–31 December 1992 (data courtesy of NOAA)

(GCM) E, of which the principal atmospheric variables used for forecasting ('prognostic') are the potential temperature, water vapor mixing ratio, and the horizontal velocity component[33]. There is of course always room for additional components or modulation in the mix: in another GCM, photochemically induced changes in stratospheric ozone reduced the impact of increased irradiance[5] by about 1/3.

Heliohydrology

Two versions of the GISS model—M20, a $4° \times 5°$ 20-layer model, and M23, the corresponding full-stratospheric model—simulate the location of the northern hemisphere storm tracks 'reasonably well' even though the number of storms is low; peak winds in the jet stream, on the other hand, are slightly high[33].

The polar front between the Hadley cells and the Ferrel cells coincides with the jet streams, undulating belts of westerly high speed winds which govern the track and vigour of the mid-latitude extratropical cyclonic systems (Fig. 7.8). The crucial issue is how they are linked to the UV flux. A plausible answer is that, besides any blanket change in the temperature of constituent gases, an inevitable latitudinal imbalance calls for the transfer of heat to middle and high latitudes, and the Hadley cell plays a key role in the transaction[31]. To this simple arrangement must be added the impact of the Coriolis effect, the apparent deflection of air masses moving over the rotating Earth.

The study of exoplanet climates, necessarily still at a preliminary stage, is especially instructive when the emphasis is put on atmospheric dynamics. Interestingly, key concepts including planetary rotation show that turbulent interactions—Feynman's bogey—may produce elongated E-W structures, including jet streams[36].

Computer modelling suggests that accentuated UV heating of stratospheric ozone may result in the weakening and broadening of the Hadley and Ferrel cells and poleward displacement of the subtropical jets[16, 35]. Other three-dimensional products of air-mass convergence and divergence are the Intertropical Convergence Zone (ITCZ), the South Pacific Convergence Zone and the Tropical Walker circulation and its El Niño-Southern Oscillation (ENSO) variants. The latitude of the ITCZ

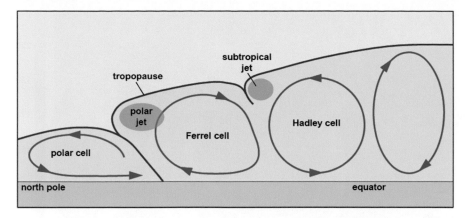

Fig. 7.8 Schematic section showing major components of the atmospheric circulation of the northern hemisphere

shifts equatorward with decreased solar activity[48]. Analysis of ENSO is generally based on models which assume that its causes are internal to the climate system[1, 4] whereas the strength of the Trade Winds is both a symptom of Hadley Cell vigour and a major factor in the east-west displacement of warm surface waters that is at the root of the El Niño cycle in the Pacific: a weakening of the Trade Winds raises the thermocline (the layer between warmer mixed water and colder deep waters) in the Western Pacific and depresses it in the Indonesian realm. Complementing the top-down UV/ozone mechanism is a proposed 'bottom-up' one which seeks to amplify a small but direct solar signal by exploiting the greater evaporation created by increased solar flux over cloud-free tropical regions[27].

Returning to the jet streams, it is evidently impossible to identify a single value for the velocity or latitude of such sinuous and mobile features (Fig. 7.9). Even so, averaging the 6-hourly maps prepared by the California Regional Weather Server for the North Atlantic polar jet, despite substantial scatter and the subjective element in identifying maxima for discontinuous features, demonstrates that daily maximum velocity, which ranges in the sample year over 79–210 knots, is at a minimum in North Hemisphere summer (Fig. 7.10). Jet stream latitude for daily maximum velocity, which in the same analysis ranges over 30°–85° N, can be determined using total column ozone[19].

Storm-track frequency[49] and the tracks themselves (Fig. 7.11) reveal shifts in latitude which are consistent with the models. The rivers fed by the cyclonic systems represent a mechanism for creating a local and tangible trace of the shifts in the circulation. Such a trace has revealed the latitudinal advance and retreat of cyclonic river regimes across the northern hemisphere middle latitudes between ~1500 and 500 year ago[25]. Once we enter the instrumental period the tracks themselves are seen to echo the two peaks in storm track latitude at the two extremes of successive

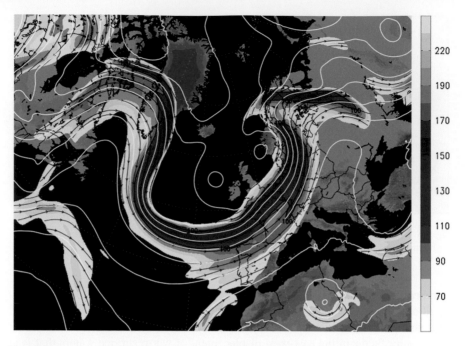

Fig. 7.9 Sample of forecast for polar-front jetstream (in mph) for North Atlantic (see Fig. 7.8). Note amplified or meridional pattern (courtesy of Guy Mears, Netweather)

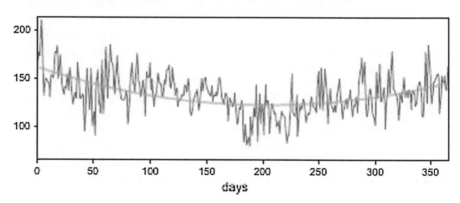

Fig. 7.10 Annual variation for 1992 of North Atlantic jet stream velocity (knots) over open ocean based on daily maps (data courtesy of California Regional Weather Server)

solar cycles. Figure 7.11 shows, for solar cycle 22 (1986–1996), the contrast between winter storm tracks near the minimum and the maximum of the cycle. Note how the tracks penetrate into the high Atlantic latitudes at solar maximum and how they are well represented in southern Europe at solar minimum.

The increase in EUV radiation that is responsible for displacing the polar jet also leaves an imprint in the radiocarbon record of tree rings and ice cores. As we have seen, when the Sun is more active the flux of GCRs, and thus the production of

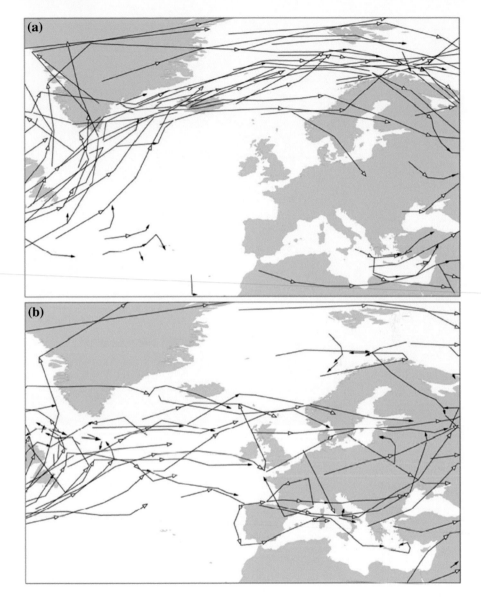

Fig. 7.11 Storm tracks for winter of 1991–2 and 1985–6, near solar maximum (**a**) and minimum (**b**) respectively, for solar cycle 22 (after NASA GISS Atlas of Extratropical Storm Tracks)

cosmogenic isotopes including ^{14}C, is reduced. In the middle latitudes of Eurasia and the Americas many natural drainage systems (that is to say channels that are free of dams and other such structures) undergo aggradation under cyclonic conditions and channel trenching at other times (Fig. 7.12a). The present-day circulation is

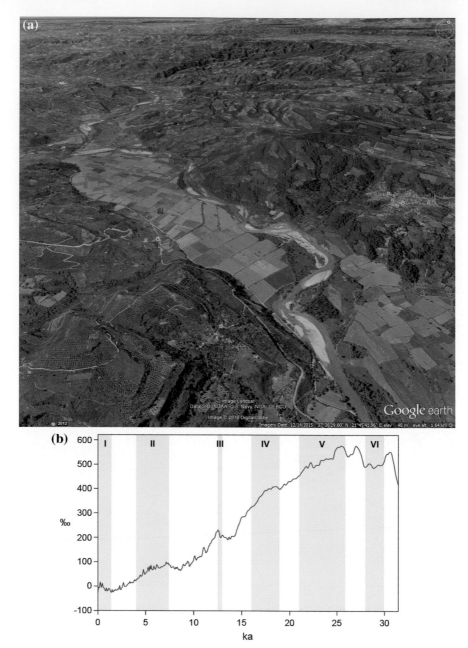

Fig. 7.12 a The Alfios valley upstream from Olympia (Google earth image). Flat-topped bottom-lands, now being trenched by shortlived streamflows, reflect a former episode of cyclonic precipitation governed by solar fluctuations. **b** Relationship between Δ^{14}C and major aggradational periods in the Mediterranean region (I–VI) in the last 30,000 years (various sources)

appended to a large-scale version of the solar cycle, namely the major dip in solar activity responsible for the Little Ice Age. In this circuitous manner fluctuations in solar UV and ^{14}C (Fig. 7.12b) supplement the short and ephemeral meterorological archive with the somewhat more durable record of river alluvium[49].

In a word, the vagaries of the Sun are writ in water[52] but immortalised by mud.

References

1. Bjerknes J (1969) Atmospheric teleconnections from the Equatorial Pacific. Month Weath Rev 97:163–172
2. Bornman JF et al (2015) Solar ultraviolet radiation and ozone depletion-driven climate change: effects on terrestrial ecosystems. Photochem Photobiol Sci 14:88–107
3. Boucher O et al (2013) Clouds and aerosols. In Stocker TF et al (eds) Climate Change 2013: The physical science basis. Cambridge Univ Press Cambridge
4. Chen D et al (2004) Predictability of El Niño over the past 148 years. Nature 428:733–736
5. Chiodo G & Polvani LM (2016) Reduction of climate sensitivity to solar forcing due to stratospheric ozone feedback. J Clim 29:4651–4663
6. Dengel S, Aeby D, Grace J (2009) A relationship between galactic cosmic radiation and tree rings. New Phyt 184:545–551
7. DeVorkin DH (1990) Defending a dream: Charles Greeley Abbot's years at the Smithsonian. J Hist Astron 21:121–136
8. Dickinson R (1975) Solar variability and the lower atmosphere. Bull Am Met Soc 56: 1246–1248
9. Diehl R et al (2016) Radioactive ^{26}Al from massive stars in the Galaxy. Nature 439:45–47
10. Ensminger PA (2001) Life under the Sun. Yale Univ Press, New Haven & London
11. Feynman R, Leighton RB, Sands ML (1963) The Feynman lectures on physics. Addison-Wesley, Reading, Mass
12. Forbush SE (1967) Solar influences on cosmic rays. Proc Nat Acad Sci 43:28–41
13. Fuller-Rowell T et al (2004) Impact of solar EUV, XUV, and X-ray variations on Earth's atmosphere. Geophys Monog 141:341–354
14. Gessler A et al (2009) Tracing carbon and oxygen isotope signals from newly assimilated sugars in the leaves to the tree-ring archive. Plant Cell Env 32:780–795
15. Gray LJ et al (2010) Solar influences on climate. Rev Geophys 48: RG4001, https://doi.org/10.1029/2009rg00028
16. Haigh JD (1999) Modelling the impact of solar variability on climate. J Atmos Solar-Terr Phys 61:63–72
17. Harrison RG, Stephenson DB (2006) Empirical evidence for a non-linear effect of galactic cosmic rays on clouds. Proc Roy Soc 462: https://doi.org/10.1098/rspa.2005.1628
18. Hood LL (1999) Effects of short-term solar UV variability on the stratosphere. J Atmos Sol-Terr Phys 61:41–51
19. Hudson RD (2012) Measurements of the movement of the jet streams at mid-latitudes, in the Northern and Southern Hemispheres, 1979 to 2010. Atmos Chem Phys 12:7797–7808
20. IPCC (1997) Introduction to simple climate models used in the IPCC second assessment report. IPCC Geneva, Switzerland
21. IPCC (2014) Climate Change 2014: Synthesis Report. IPCC, Geneva, Switzerland

22. Kirkby J et al (2011) Role of sulphuric acid, ammonia and galactic cosmic rays in atmospheric aerosol nucleation. Nature 476:429–433
23. Laken BA, Kniveton DR, Frogley MR (2010) Cosmic rays linked to rapid mid-latitude cloud changes. Atmos Chem Phys 10:10941–10948
24. Lam MM, Chisham G, Freeman MP (2014) Solar wind-driven geopotential height anomalies originate in the Antarctic lower troposphere. Geophys Res Lett 41:6509–6514
25. Leopold LB, Vita-Finzi C (1998) Valley changes in the Mediterranean and America. Proc Am Phil Soc 142:1–17
26. Li KJ et al (2012) Why is the solar constant not a constant? Astrophys J 747:135
27. Meehl GA et al (2009) Amplifying the Pacific climate system response to a small 11 year solar cycle forcing. Science 325:1114–1118
28. Myhre G et al (2013) Climate Change 2013: The Physical Science Basis. Cambridge Univ Press, Cambridge
29. Ney ER (1959) Cosmic radiation and the weather. Nature 183:451–452
30. Rozema J et al (2009) UV-B absorbing compounds in present-day and fossil pollen, spores, cuticles, seed coats and wood: evaluation of a proxy for solar UV radiation. Photochem Photobiol Sci 8:1233–1243S
31. Salby ML (1996) Fundamentals of Atmospheric Physics. Academic, San Diego
32. Salby ML, Callaghan PF (2004) Evidence of the solar cycle in the general circulation of the stratosphere. J Clim 17:34–46
33. Schmidt GA et al (2006) Present-day atmospheric simulations using GISS Model E: comparison to in situ, satellite, and reanalysis data. J Clim 19:153–192
34. Shaviv NJ (2002) Cosmic ray diffusion from the Galactic Spiral Arms, iron meteorites, and a possible climatic connection? Phys Rev Lett 89:051102
35. Shindell D (1999) Solar cycle variability, ozone, and climate. Science 284:305–308
36. Showman AP, Cho J Y-K, Menou K (2009) Atmospheric circulation of exoplanets. In Sweager S (ed) Exoplanets, Univ Arizona Press, Tucson, 471–516
37. SNSF (Swiss National Science Foundation) (2017) Future and past solar influence on the terrestrial climate. Geneva
38. Soudant A et al (2016) Intra-annual variability of wood formation and d¹³C in tree-rings at Hyytiälä, Finland. Agric Forest Met 224:17–29
39. Sturrock PA (2009) Combined analysis of solar neutrino and solar irradiance data: further evidence for variability of the solar neutrino flux and its implications concerning the solar core
40. Svensmark H, Friis-Christensen E (1997) Variation of cosmic ray flux and global cloud coverage—a missing link in solar-climate relationships. J Atmos Sol-Terr Phys 59:1225–1232
41. Svensmark et al (2016) The response of clouds and aerosols to cosmic ray decreases. Jour Geophys Res, Space Phys 121:8152–8181
42. Svensmark H et al (2017) Increased ionization supports growth of aerosols into cloud condensation nuclei. Nature comm 8: 2199
43. Takahashi K, Tokimitsu Y, Yassue K (2005) Climatic factors affecting the tree-ring width of *Betula ermanii* at the timberline of Mount Norikura, central Japan. Ecol Res 20:445–451
44. Takahashi Y et al (2009) 27-day variation in cloud amount and relationship to the solar cycle. Atmos Chem Phys Disc 9:15327–15338
45. Thayer JP et al (2008) Thermospheric density oscillations due to periodic solar wind high-speed streams.J. Geophys Res 113:A06307
46. Thompson MJ et al (2003) The internal rotation of the Sun. Ann Rev Astron Astrophys 41:599–643
47. Tinsley BA (2008) The global atmospheric electric circuit and its effects on cloud microphysics. Rep Prog Phys 71:066801
48. Versteegh GJM (2005) Solar forcing of climate. 2: evidence from the past. Space Sci Rev 120:243–286
49. Vita-Finzi C (2008) Fluvial solar signals. Geol Soc Lond Spec Pub 296:105–115
50. Vita-Finzi C (2010) The Dicke cycle: a 27-day solar oscillation. J Atmos Sol-Terr Phys 72:139–142

51. Vita-Finzi C (2014) Towards a solar system timescale. Astron Geophys 55:4.27–4.29
52. The poet Keats' epitaph, according to Lord Houghton, read 'Here lies one whose name was writ in water'

Index

© Springer Nature Switzerland AG 2018
C. Vita-Finzi, *The Sun Today*, https://doi.org/10.1007/978-3-030-04079-6

Printed in the United States
By Bookmasters